APPLIED REGRESSION ANALYSIS AND EXPERIMENTAL DESIGN

STATISTICS: Textbooks and Monographs

A SERIES EDITED BY

D. B. OWEN, Coordinating Editor

Department of Statistics
Southern Methodist University
Dallas, Texas

Vol. 1: The Generalized Jackknife Statistic, *H. L. Gray and W. R. Schucany*
Vol. 2: Multivariate Analysis, *Anant M. Kshirsagar*
Vol. 3: Statistics and Society, *Walter T. Federer*
Vol. 4: Multivariate Analysis: A Selected and Abstracted Bibliography, 1957-1972, *Kocherlakota Subrahmaniam and Kathleen Subrahmaniam* (out of print)
Vol. 5: Design of Experiments: A Realistic Approach, *Virgil L. Anderson and Robert A. McLean*
Vol. 6: Statistical and Mathematical Aspects of Pollution Problems, *John W. Pratt*
Vol. 7: Introduction to Probability and Statistics (in two parts), Part I: Probability; Part II: Statistics, *Narayan C. Giri*
Vol. 8: Statistical Theory of the Analysis of Experimental Designs, *J. Ogawa*
Vol. 9: Statistical Techniques in Simulation (in two parts), *Jack P. C. Kleijnen*
Vol. 10: Data Quality Control and Editing, *Joseph I. Naus* (out of print)
Vol. 11: Cost of Living Index Numbers: Practice, Precision, and Theory, *Kali S. Banerjee*
Vol. 12: Weighing Designs: For Chemistry, Medicine, Economics, Operations Research, Statistics, *Kali S. Banerjee*
Vol. 13: The Search for Oil: Some Statistical Methods and Techniques, *edited by D. B. Owen*
Vol. 14: Sample Size Choice: Charts for Experiments with Linear Models, *Robert E. Odeh and Martin Fox*
Vol. 15: Statistical Methods for Engineers and Scientists, *Robert M. Bethea, Benjamin S. Duran, and Thomas L. Boullion*
Vol. 16: Statistical Quality Control Methods, *Irving W. Burr*
Vol. 17: On the History of Statistics and Probability, *edited by D. B. Owen*
Vol. 18: Econometrics, *Peter Schmidt*
Vol. 19: Sufficient Statistics: Selected Contributions, *Vasant S. Huzurbazar (edited by Anant M. Kshirsagar)*
Vol. 20: Handbook of Statistical Distributions, *Jagdish K. Patel, C. H. Kapadia, and D. B. Owen*
Vol. 21: Case Studies in Sample Design, *A. C. Rosander*
Vol. 22: Pocket Book of Statistical Tables, *compiled by R. E. Odeh, D. B. Owen, Z. W. Birnbaum, and L. Fisher*
Vol. 23: The Information in Contingency Tables, *D. V. Gokhale and Solomon Kullback*
Vol. 24: Statistical Analysis of Reliability and Life-Testing Models: Theory and Methods, *Lee J. Bain*
Vol. 25: Elementary Statistical Quality Control, *Irving W. Burr*
Vol. 26: An Introduction to Probability and Statistics Using BASIC, *Richard A. Groeneveld*
Vol. 27: Basic Applied Statistics, *B. L. Raktoe and J. J. Hubert*
Vol. 28: A Primer in Probability, *Kathleen Subrahmaniam*
Vol. 29: Random Processes: A First Look, *R. Syski*

Vol. 30: Regression Methods: A Tool for Data Analysis, *Rudolf J. Freund and Paul D. Minton*
Vol. 31: Randomization Tests, *Eugene S. Edgington*
Vol. 32: Tables for Normal Tolerance Limits, Sampling Plans, and Screening, *Robert E. Odeh and D. B. Owen*
Vol. 33: Statistical Computing, *William J. Kennedy, Jr. and James E. Gentle*
Vol. 34: Regression Analysis and Its Application: A Data-Oriented Approach, *Richard F. Gunst and Robert L. Mason*
Vol. 35: Scientific Strategies to Save Your Life, *I. D. J. Bross*
Vol. 36: Statistics in the Pharmaceutical Industry, *edited by C. Ralph Buncher and Jia-Yeong Tsay*
Vol. 37: Sampling from a Finite Population, *J. Hajek*
Vol. 38: Statistical Modeling Techniques, *S. S. Shapiro*
Vol. 39: Statistical Theory and Inference in Research, *T. A. Bancroft and C.-P. Han*
Vol. 40: Handbook of the Normal Distribution, *Jagdish K. Patel and Campbell B. Read*
Vol. 41: Recent Advances in Regression Methods, *Hrishikesh D. Vinod and Aman Ullah*
Vol. 42: Acceptance Sampling in Quality Control, *Edward G. Schilling*
Vol. 43: The Randomized Clinical Trial and Therapeutic Decisions, *edited by Niels Tygstrup, John M. Lachin, and Erik Juhl*
Vol. 44: Regression Analysis of Survival Data in Cancer Chemotherapy, *Walter H. Carter, Jr., Galen L. Wampler, and Donald M. Stablein*
Vol. 45: A Course in Linear Models, *Anant M. Kshirsagar*
Vol. 46: Clinical Trials: Issues and Approaches, *edited by Stanley H. Shapiro and Thomas H. Louis*
Vol. 47: Statistical Analysis of DNA Sequence Data, *edited by B. S. Weir*
Vol. 48: Nonlinear Regression Modeling: A Unified Practical Approach, *David A. Ratkowsky*
Vol. 49: Attribute Sampling Plans, Tables of Tests and Confidence Limits for Proportions, *Robert E. Odeh and D. B. Owen*
Vol. 50: Experimental Design, Statistical Models, and Genetic Statistics, *edited by Klaus Hinkelmann*
Vol. 51: Statistical Methods for Cancer Studies, *edited by Richard G. Cornell*
Vol. 52: Practical Statistical Sampling for Auditors, *Arthur J. Wilburn*
Vol. 53: Statistical Signal Processing, *edited by Edward J. Wegman and James G. Smith*
Vol. 54: Self-Organizing Methods in Modeling: GMDH Type Algorithms, *edited by Stanley J. Farlow*
Vol. 55: Applied Factorial and Fractional Designs, *Robert A. McLean and Virgil L. Anderson*
Vol. 56: Design of Experiments: Ranking and Selection, *edited by Thomas J. Santner and Ajit C. Tamhane*
Vol. 57: Statistical Methods for Engineers and Scientists. Second Edition, Revised and Expanded, *Robert M. Bethea, Benjamin S. Duran, and Thomas L. Boullion*
Vol. 58: Ensemble Modeling: Inference from Small-Scale Properties to Large-Scale Systems, *Alan E. Gelfand and Crayton C. Walker*
Vol. 59: Computer Modeling for Business and Industry, *Bruce L. Bowerman and Richard T. O'Connell*
Vol. 60: Bayesian Analysis of Linear Models, *Lyle D. Broemeling*
Vol. 61: Methodological Issues for Health Care Surveys, *Brenda Cox and Steven Cohen*
Vol. 62: Applied Regression Analysis and Experimental Design, *Richard J. Brook and Gregory C. Arnold*

OTHER VOLUMES IN PREPARATION

APPLIED REGRESSION ANALYSIS AND EXPERIMENTAL DESIGN

RICHARD J. BROOK
GREGORY C. ARNOLD

Department of Mathematics and Statistics
Massey University
Palmerston North, New Zealand

MARCEL DEKKER, INC. New York and Basel

Library of Congress Cataloging in Publication Data

Brook, Richard J.
Applied regression analysis and experimental design.

(Statistics, textbooks and monographs ; vol. 62)
Includes index.
1. Regression analysis. 2. Experimental design.
I. Arnold, G. C. (Gregory C.), [date] . II. Title.
III. Series: Statistics, textbooks and monographs ; v. 62.
QA278.2.B76 1985 519.5'36 85-4361
ISBN 0-8247-7252-0

MARCEL DEKKER, INC.
270 Madison Avenue, New York, New York 10016

Current printing (last digit):
10 9 8 7 6 5 4 3 2 1

PRINTED IN THE UNITED STATES OF AMERICA

PREFACE

This textbook was written to provide a clear and concise discussion of regression and experimental design models. Equal weighting is given to both of these important topics which are applicable, respectively, to observational data and data collected in a controlled manner. The unifying concepts for these topics are those of linear models so that the principles and applications of such models are considered in some detail.

We have assumed that the reader will have had some exposure to the basic ideas of statistical theory and practice as well as some grounding in linear algebra. Consequently, this text will be found useful in undergraduate/graduate courses as well as being of interest to a wider audience, including numerate practitioners.

We felt that it was important to consider variables, which can be written as columns of data, as geometric vectors. Behind the vector notation is always a geometric picture which we believe helps to make the results intuitively plausible without requiring an excess of theory. In this way we have tried to give readers an understanding of the value and purpose of the methods described, so that the book is not about the theory of linear models, but their applications. To this end, we have included an appendix containing seven data sets. These are referred to frequently throughout the book and they form the basis for many of the problems given at the end of each chapter.

We assume that the reader will have computer packages available. We have not considered in any detail the problems of numerical analysis or the methods of computation. Instead we have discussed the strengths, weaknesses and ambiguities of computer output. For the reader, this means that space-consuming descriptions of computations are kept to a minimum.

We have concentrated on the traditional least squares method but we point out its possible weaknesses and indicate why more recent sophisticated techniques are being explored.

We have included such topics as subset selection procedures, randomization, and blocking. It is our hope that students, having been introduced to these ideas in the general context of the linear model, will be well equipped to pick up the details they need for their future work from more specialised texts.

In the first four chapters, we cover the linear model in the regression context. We consider topics of how to fit a line, how to test whether it is a good fit, variable selection, and how to identify and cope with peculiar values. In the remaining four chapters we turn to experimental design, and consider the problem of constructing and estimating meaningful functions of treatment parameters, of utilising structure in the experimental units as blocks, and of fitting the two together to give a useful experiment.

This book represents the final version of course notes which have evolved over several years. We would like to thank our students for their patience as the course notes were corrected and improved. We acknowledge the value of their comments and less tangible reactions. Our data sets and examples, with varying degrees of modification, have many sources, but we particularly thank John Baker, Selwyn Jebson, David Johns, Mike O'Callaghan and Ken Ryba of Massey University, Dr R. M. Gous of the University of Natal, and Julie Anderson of the New Zealand Dairy Research Institute for giving us access to a wide range of data.

Richard J. Brook
Gregory C. Arnold

CONTENTS

APPLIED REGRESSION ANALYSIS AND EXPERIMENTAL DESIGN

1

FITTING A MODEL TO DATA

1.1 INTRODUCTION

The title of this chapter could well be the title of this book. In the first four chapters, we consider problems associated with fitting a regression model and in the last four we consider experimental designs. Mathematically, the two topics use the same model. The term regression is used when the model is fitted to observational data, and experimental design is used when the data is carefully organized to give the model special properties. For some data, the distinction may not be at all clear or, indeed, relevant. We shall consider sets of data consisting of observations of a variable of interest which we shall call y, and we shall assume that these observations are a random sample from a population, usually infinite, of possible values. It is this population which is of primary interest, and not the sample, for in trying to fit models to the data we are really trying to fit models to the population from which the sample is drawn. For each observation, y, the model will be of the form

$$\text{observed } y = \text{population mean} + \text{deviation} \qquad (1.1.1)$$

The population mean may depend on the corresponding values of a predictor variable which we often label as x. For this reason, y is called the dependent variable. The deviation term indicates the individual peculiarity of the observation, y, which makes it differ from the population mean.

As an example, \$y could be the price paid for a house in a certain city. The population mean could be thought of as the mean price paid for houses in that city, presumably in a given time period. In this case the deviation term could be very large as house prices would vary greatly depending on a number of factors such as the size and condition of the house as well as its position in the city. In New Zealand, each house is given a government valuation, GV, which is reconsidered on a five year cycle. The price paid for a house will depend to some extent on its GV. The regression model could then be written in terms of \$x, the GV, as:

$$\underset{\text{price}}{y} = \underset{\text{population mean}}{\alpha + \beta x} + \underset{\text{deviation}}{\varepsilon} \tag{1.1.2}$$

As the population mean is now written as a function of the GV, the deviations will tend to be smaller. Figure 1.1.1 indicates possible values of y when x=20,000 and x=50,000. Theoretically, all values of y may be possible for each value of x but, in practice, the y values would be reasonably close to the value representing the population mean.

The model could easily be extended by adding other predictor variables such as the age of the house or its size. Each deviation

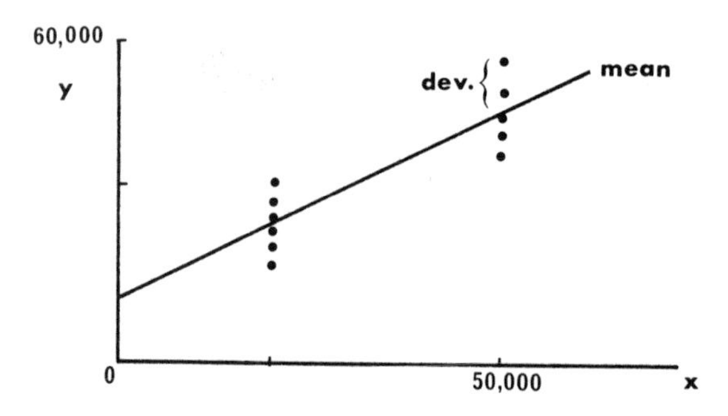

FIGURE 1.1.1 House prices, y, regressed against GV, x.

term would tend to be smaller now as the population mean accounts for the variation in prices due to these additional variables. The deviation term can be thought of as accounting for the variations in prices unexplained by the mean.

Another example, this time from horticulture, would be a model in which y is the yield, in kilograms, of apples per hectare for different orchards. The population mean could be written as a function of the amount of fertilizer added, the amount of insecticide spray used, and the rainfall. In this case, the deviation term would include unexplained physical factors such as varying fertility of the soils as well as possible errors of measurement in weighing the apples.

In each of these examples, a model is postulated and as it relates to the population, of which we know only the small amount of information provided by the sample, then we must use some method of deciding which part of y relates to the population mean and which to the deviation. We shall use the method of least squares to do this.

1.2 HOW TO FIT A LINE

1.2.1 The Method of Least Squares

As the deviation term involves the unexplained variation in y, we try to minimise this in some way. Suppose we postulate that the mean value of y is a function of x. That is

$$E(y) = f(x)$$

Then for a sample of n pairs of y's with their corresponding x's we have

$$\underset{\text{observed } y}{y_i} = \underset{\text{mean of } y}{f(x_i)} + \underset{\text{deviation}}{\varepsilon_i} \qquad 1 \leqq i \leqq n \qquad (1.2.1)$$

The above notation assumes that the x's are not random variables but are fixed in advance. If the x's were in fact random variables we should write

$$f(x_i) = E(y_i \mid X_i = x_i)$$
$$= \text{mean of } Y_i \text{ given that } X_i = x_i$$

which gives the same results. We will therefore assume in future that the x's are fixed.

The simplest example of a function f would arise if y was proportional to x. We could imagine a situation where an inspector of weights and measures set out to test the scales used by shopkeepers. In this case, the x's would be the weights of standard measures while y's would be the corresponding weights indicated by the shopkeeper's scales. The model would be

y_i	=	βx_i	+	ε_i	
weight shown by scales		parameter standard measure		deviation	(1.2.2)

The mean value of y when $x = x_i$ is given by

$$E(y_i) = \beta x_i = f(x_i) \tag{1.2.3}$$

This is called a regression curve. In this simple example we would expect the parameter β to be 1, or at least close to 1. We think of the parameters as being fixed numbers which describe some attributes of the population.

The readings of the scales, the y's, will fluctuate, some being above the mean, f(x), in which case the deviation, ε, will be positive while others will be below the mean and the corresponding ε will be negative.

The method of least squares uses the sample of n values of x and y to estimate population parameters by minimizing the deviations ε. More specifically, we seek a value of β which we will label b to minimize the sum of squares of the ε_i, that is

$$S = \sum_{i=1}^{n} \varepsilon_i^2 = \sum_{i=1}^{n} [y_i - f(x_i)]^2 \tag{1.2.4}$$

If the mean, $f(x)$, has the simple structure of the model (1.2.2)

$$S = \sum_{i=1}^{n} [y_i - \beta x_i]^2 \tag{1.2.5}$$

Methods of algebra or calculus can be employed to yield

$$\sum_{i=1}^{n} [y_i - b x_i] x_i = 0 \tag{1.2.6}$$

Rearranging (1.2.6), the least squares estimate of β is the value b which solves the equation

$$\sum_{1=1}^{n} b x_i^2 = \sum_{i=1}^{n} x_i y_i$$

$$\text{or} \quad b = \Sigma x_i y_i / \Sigma x_i^2 \tag{1.2.7}$$

This equation is called the <u>normal equation</u>. For those who appreciate calculus, it could be noted that this equation (1.2.7) can also be written as

$$\sum [y_i - f(x_i)] \frac{\partial f}{\partial \beta} = 0 \tag{1.2.8}$$

where $\frac{\partial f}{\partial \beta}$ is the partial derivative of $f(x;\beta)$ with respect to β. For this simple model without a constant, we have:

$$\begin{aligned} &\text{the regression curve is} \quad E(y_i) = f(x_i) = \beta x_i \\ &\text{and the estimate of it is} \quad \hat{y}_i = \hat{f}(x_i) = b x_i \end{aligned} \tag{1.2.9}$$

Equation 1.2.9 is called the <u>prediction curve</u>. Notice that:

(i) $\hat{y}_i$ estimates the mean value of y when $x = x_i$.
(ii) The difference $y_i - \hat{y}_i = e_i$, which is called the residual.

(iii) Parameters are written as Greek letters.
(iv) Estimates of the parameters are written in Roman letters.

Even with the simple problem of calibration of scales it may be sensible to add an intercept term into the model for it may be conceivable that all the scales weigh consistently on the high side by an amount α. The model is then

$$y_i = \alpha + \beta x_i + \varepsilon_i \tag{1.2.10}$$

The normal equations become

$$\sum [y_i - f(x_i)] \frac{\partial f}{\partial \alpha} = 0$$

$$\sum [y_i - f(x_i)] \frac{\partial f}{\partial \beta} = 0 \tag{1.2.11}$$

From (1.2.11), or using algebra, and noting that $\Sigma a_i = na$, we obtain

$$a\,n + b\,\Sigma\,x_i = \Sigma\,y_i$$

$$a\,\Sigma\,x_i + b\,\Sigma\,x_i^2 = \Sigma\,x_i y_i \tag{1.2.12}$$

Elementary texts give the solution of these normal equations as

$$b = \left[\Sigma\,(x_i-\bar{x})(y_i-\bar{y})\right]/\left[\Sigma\,(x_i-\bar{x})^2\right] \tag{1.2.13}$$

$$a = \bar{y} - b\bar{x}$$

Here, $\bar{x}$ and $\bar{y}$ are the sample means.

It is easy to extend (1.2.12) to many variables. For a model with k variables we need to use double subscripts as follows

$$y_i = \beta_0 x_{i0} + \beta_1 x_{i1} + \cdots + \beta_k x_{ik} + \varepsilon_i$$

where $x_{i0} = 1$ if an intercept term is included. The normal equations are

	C0	C1		Ck	Cy
R0	$b_0 \sum x_{i0}^2$	$+ b_1 \sum x_{i0} x_{i1}$	$+ \cdots +$	$b_k \sum x_{i0} x_{ik}$	$= \sum x_{i0} y_i$
R1	$b_0 \sum x_{i1} x_{i0}$	$+ b_1 \sum x_{i1}^2$	$+ \cdots +$	$b_k \sum x_{i1} x_{ik}$	$= \sum x_{i1} y_i$
	$\vdots$	$\vdots$		$\vdots$	$\vdots$
Rk	$b_0 \sum x_{ik} x_{i0}$	$+ b1 \sum x_{ik} x_{i1}$	$+ \cdots +$	$b_k \sum x_{ik}^2$	$= \sum x_{ik} y_i$ (1.2.14)

Notice that R0 (Row 0) involves x_0 in every term and in general Rj involves x_j, which is analagous to (1.2.11) with the derivative taken with respect to β_j. Similarly C0 (Col 0) involves x_0 in every term, and in general Cj involves x_j and Cy involves y in every term.

Example 1.2.1

Consider the simple example of the calibrating of scales where x kg is the "true" weight and y kg the weight indicated by a certain scale. The values of x and y are given in Table 1.2.1. For the model without an intercept term

$$\hat{y} = bx = 0.97 x \quad \text{from (1.2.7)}$$

If an intercept term is included, the normal equations of (1.2.12) become

$$5.0 a + 7.5 b = 7.55$$
$$7.5 a + 13.75 b = 13.375$$

TABLE 1.2.1 Scale Calibration Data

y	x
0.70	0.5
1.15	1.0
1.35	1.5
2.05	2.0
2.30	2.5
----	---
$\Sigma y = 7.55$	$\Sigma x = 7.5$
$\Sigma xy = 13.375$	$\Sigma x^2 = 13.75$

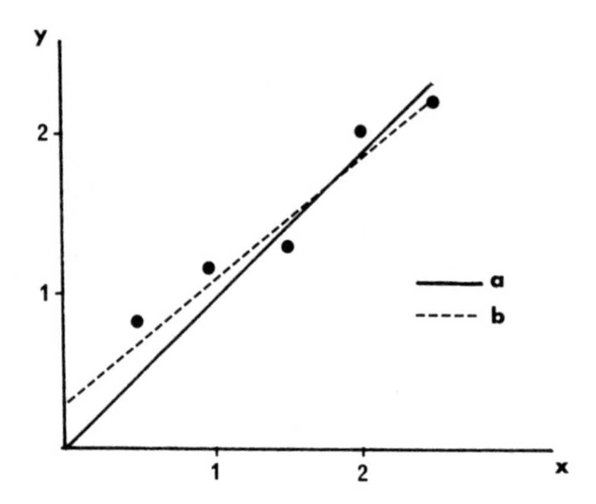

FIGURE 1.2.1 Prediction curves. a: with intercept, b: no intercept.

The solution to these equations is a = 0.28, b = 0.82 giving the prediction curve

$$\hat{y} = 0.28 + 0.82\ x$$

The prediction curves are shown in Figure 1.2.1.

1.2.2 The Assumptions of Least Squares

We have used the method of least squares without considering assumptions on the model. It is usual, however, to make certain assumptions which justify the use of the least squares approach. In particular, the estimates and predicted values we obtain will be optimal in the sense of being unbiased and having the smallest variance among all unbiased linear estimates provided that the following four assumptions hold:

(i) The x values are fixed and not random variables
(ii) The deviations are independent
(iii) The deviations have a mean of zero and
(iv) The variance of the deviations is constant and does not depend on (say) the x values.

If we add a fifth assumption, namely,

(v) The deviations are normally distributed,

then the estimates of the parameters are the same as would be obtained from maximum likelihood, which gives us further theoretical assurances. For the development followed in this book, we are more concerned that this property ensures that estimates of parameters and predicted values of y are also distributed normally leading to F-tests and confidence intervals based on the t-statistics. In fact, means, normality and the method of least squares go hand in hand. It is not very surprising that least squares is an optimal approach if the above assumptions are true.

1.2.3 Other Ways of Fitting a Curve

The main problem with the approach of least squares is that a large deviation will have an even larger square and this deviation may have an unduly large influence on the fitted curve. To guard against such distortions we could try to isolate large deviations. We consider this in more detail in Chapter 4 under outliers and sensitive points. Alternatively, we could seek estimates which minimize a different function of the deviations.

If the model is expressed in terms of the population median of y, rather than its mean, another method of fitting a curve would be by minimizing T, the sum of the absolute values of deviations, that is

$$T = \sum_{i=1}^{n} | \varepsilon_i |$$

Although this is a sensible approach which works well, the actual mathematics is difficult when the distributions of estimates are sought. Hogg (1974) suggests minimizing

$$T = \sum | \varepsilon_i |^p \quad \text{with} \quad 1<p<2$$

and p = 1.5, in particular, may be a reasonable compromise. Again it is difficult to determine the exact distributions of the resulting estimates. If we are not so much interested in testing hypothesis as estimating coefficients then this method provides estimates which are robust in the sense that they are not unduly affected by large deviations.

Notice that the deviations are the vertical distances from the regression line. It might, perhaps, seem more logical, or at least more symmetrical, to consider the perpendicular (orthogonal) distances from the regression line. However when our major concern is predicting y from x the vertical distances are more relevant because they represent the prediction error.

1.3 RESIDUALS

We would expect the residuals from the sample to mirror in some way the deviations of the model. Under the assumptions we have made that the deviations are distributed normally and independently with mean zero and constant variance, the graph of the deviations against the population mean would fall in a horizontal band.

The possibility always exists that a few deviations may be very large or very small. We expect, though, that the deviations will be grouped about zero and spread out as far for small x as they are for large x. Likewise in the sample, we expect that a graph of the residuals against the predicted values would fall in a horizontal band. If the residuals do not fall into such a band, we must query our assumptions. A few types of plots are indicated by Figure 1.3.1. The first two suggest that either a transformation of the y variable or the inclusion of a quadratic predictor term may be appropriate.

Other residual plots may be of value in deciding the form of the model, how well it fits, particular trends in the data or suspect points. In particular the following should be considered:

(i) Residual plots against predictors in the model. A curved relationship may suggest that quadratic or high powers of the predictor should be included.

(ii) Residual plots against possible predictors, not already in model. Linear or quadratic trends, for example should be noted.

(iii) Residual plots against time. The prediction equation may only be appropriate over part of the range. For example, economic data may follow a certain relationship up to some

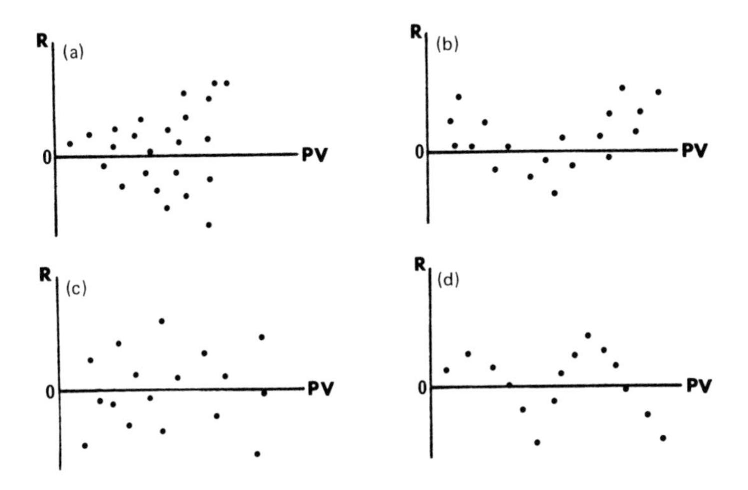

FIGURE 1.3.1 Some types of residual plots. a. Variance increasing - try a log transformation. b. Curvature - include a quadratic term. c. No problems apparent. d. Deviations not independent - try weighted least squares. R = residuals, PV = predicted values.

point in time at which government policy changed customs duty or subsidies. In some contexts time could be replaced by geographical area if observations were taken linearly.

We should also be alert to the possibility that the model may fit quite well for a certain range of predictor variables and a plot of the residuals may reveal this. Lack of independence of the deviations may be revealed by a cyclical pattern of residuals. We shall have more to say on this in Chapter 2, Section 8.2

Example 1.3.1

The predicted values of Example 1.2.1 can be formed from the prediction curve and the residuals from the prediction equation.

x	y	$\hat{y} = a+bx$	$e = y-\hat{y}$	$x\,e$
0.5	0.70	0.69	0.01	0.005
1.0	1.15	1.10	0.05	0.050
1.5	1.35	1.51	-0.16	-0.240
2.0	2.05	1.92	0.13	0.260
2.5	2.30	2.33	-0.03	-0.075
			$\Sigma e = 0.0$	$\Sigma xe = 0.0$

Notice that, in this example, the sum of the residuals is zero. This follows from the fact that the sum of the observed y = the sum of the predicted y. For the model we are considering, which has an intercept term, we can see that this fact follows from the first normal equation of (1.2.12). From the second normal equation we see that

$$\Sigma x_i \hat{y}_i = \Sigma x_i y_i \quad \text{or} \quad \Sigma x_i e_i = 0$$

For the model without an intercept term, the normal equation of (1.2.7) should show that $\Sigma x_i e_i = 0$ but Σe_i is not zero in general.

The residuals can be used to estimate variance of the deviations. The estimate we take is the sample variance of the residuals, namely:

$$\begin{aligned} s^2 &= \sum e_i^2/(n-2) \\ &= \sum (y_i - \hat{y}_i)^2/(n-2) \end{aligned}$$

The denominator is the number of degrees of freedom and it is found by (# points - # parameters to be estimated). Alternatively, we can see that there are two restrictions on the residuals, namely $\Sigma e_i = \Sigma x_i e_i = 0$. The estimate of the variance is:

$$s^2 = \sum e_i^2/(n-2) = 0.046 / 3 = 0.0153$$

The estimated standard error is $s = 0.124$.

1.4 TRANSFORMATIONS TO OBTAIN LINEARITY

Two variables, x and y, may be closely related but the relationship may not be linear. Ideally, theoretical clues would be present which point to a particular relationship such as an exponential growth model which is common in biology. Without such clues, we could firstly examine a scatter plot of y against x.

Sometimes we may recognize a mathematical model which fits the data well. Otherwise, we try to choose a simple transformation such

TABLE 1.4.1. Common Power Transformations

p	Name	Effect
-	exponential	stretches
3	cube	large
2	square	values
1	"raw"	
0.5	square root	shrinks
0	logarithmic	large
-0.5	reciprocal root	values
-1	reciprocal	

as raising the variable to a power p as in Table 1.4.1. A power of 1 leaves the variable unchanged, that is as raw data. As we proceed up or down the table from 1, the strength of the transformation increases; as we move up the table the transformation stretches larger values relatively more than smaller ones. Although the exponential does not fit in very well, we have included it as it is the inverse of the logarithmic transformation. Other fractional powers could be used but they may be difficult to interpret.

It would be feasible to transform either y or x, and, indeed, a transformation of y would be equivalent to the inverse transformation of x. For example, squaring y would be equivalent to taking the square root of x. If there are two or more predictor variables, it is often advisable to transform these in different ways rather than y, for if y is transformed to be linearly related to one predictor variable it may then not be linearly related to another.

In Figure 1.4.1, it is clear that we should stretch out the graph by increasing large x values , or, alternatively, reduce large y values. Thus, we could try changing x to x-squared, or y to the square root of y. One point to be kept in mind here is that for $p>0$, $y=0$ when $x=0$ so that it may be advisable before invoking the power transformation to change the origin; in particular,we could change y to $(y-a)$, and a good guess may be $a = 30$. In Figure 1.4.2, it seems that large x values and large y values should be reduced suggesting that a reciprocal transformation may be appropriate. This would require the x and y axes to be asymptotes which, in particular, would

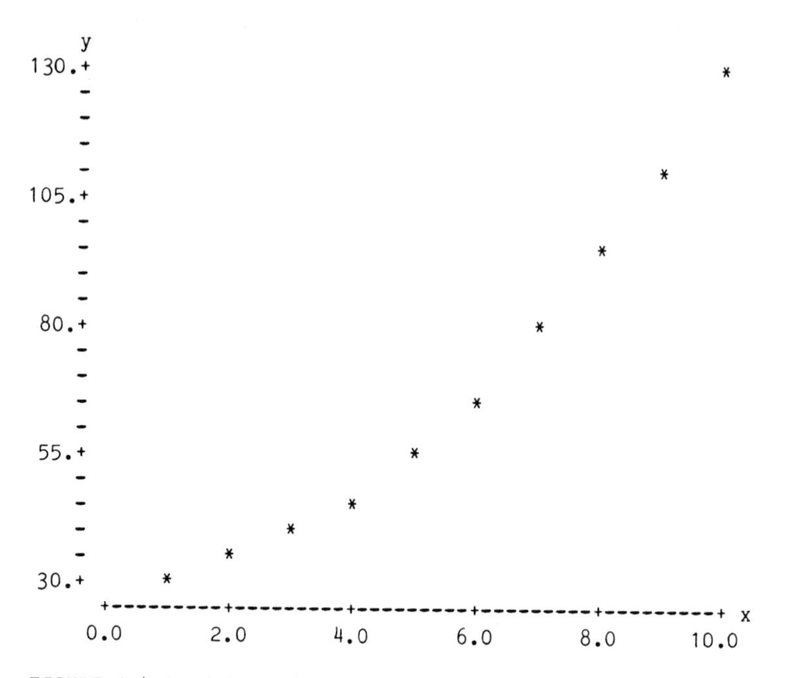

FIGURE 1.4.1 A transformation is required

mean that a constant, perhaps 14, should be subtracted from y. We could try changing

$$y \text{ to } (y - 14)^{-1}$$
$$\text{or } y \text{ to } (y - 14)^{-0.5}$$
$$\text{or } x \text{ to } x^{-1}$$
$$\text{or } x \text{ to } x^{-0.5}$$

None of these work particularly well. A logarithmic transformation changing x to log x is better, as it should be for this artificial data set was constructed from such a model. The result of this transformation is shown in Figure 1.4.3. Clearly, finding a suitable transformation may require considerable trial-and-error attempts and a number of possibilities need to be kept in mind. Besides those already mentioned, it may be that different models may

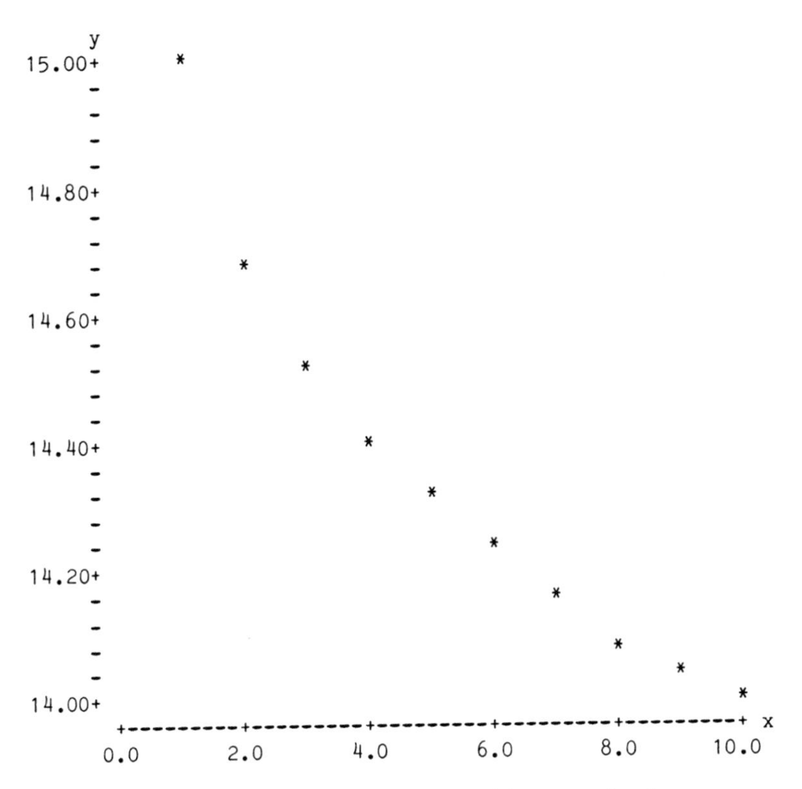

FIGURE 1.4.2 A reciprocal transformation required

fit different ranges of x; for example,for small values of x the graph may be linear but for larger x the graph may be curved suggesting that a quadratic model would be better.

In a later section we consider the effects of non-constant variance of y and the transformations which may be appropriate. Here, we have only considered transformations to obtain a linear fit.

The effectiveness of a transformation can be examined by fitting a line to the transformed values and plotting the residuals against the predicted values and also against predictor variables. The residuals should fit approximately into a horizontal band with no obvious trends being evident.

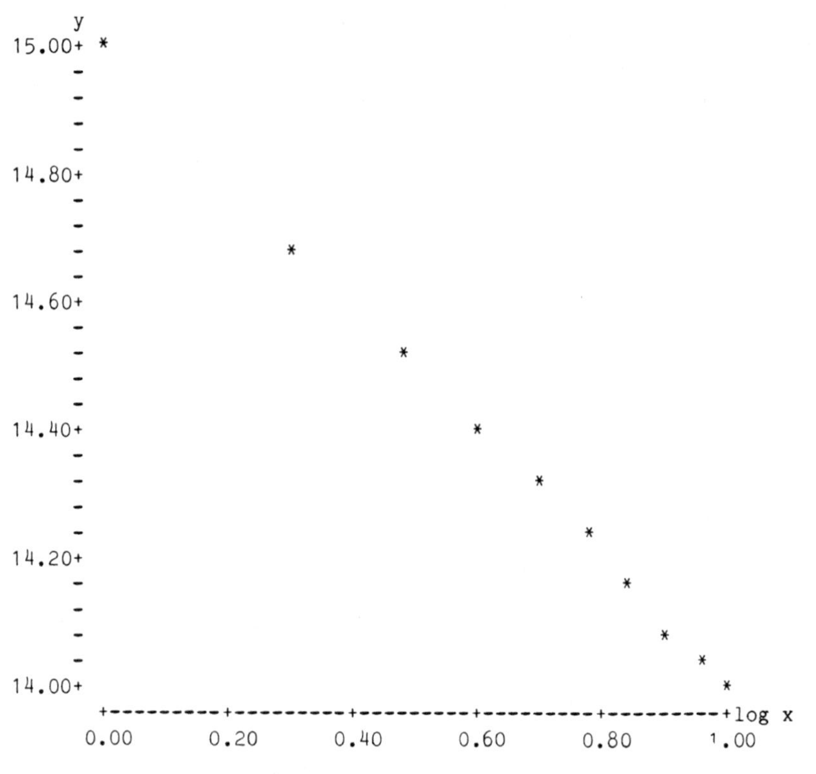

FIGURE 1.4.3 A log transformation applied.

1.5 FITTING A MODEL USING VECTORS AND MATRICES

Appendix A contains a review of vectors, vector spaces and matrices and some readers may wish to refer to that section while reading the following.

In regression, we consider the relationship between a string of values of the dependent variable, y, and one or more strings of corresponding values of the predictor variables, the x's. It is useful to think of each string as a vector for it turns out that the relationships of interest between the variables are encapsulated in the lengths of the vectors and the angles between them. For the simple Example 1.3.1, the x and y readings can be written as column vectors:

$$\mathbf{x} = \begin{vmatrix} 0.5 \\ 1.0 \\ 1.5 \\ 2.0 \\ 2.5 \end{vmatrix} \qquad \mathbf{y} = \begin{vmatrix} 0.70 \\ 1.15 \\ 1.35 \\ 2.05 \\ 2.30 \end{vmatrix}$$

$$\Sigma x_i^2 = \mathbf{x}^T \mathbf{x} = 13.75 \qquad \Sigma x_i y_i = \mathbf{x}^T \mathbf{y} = 13.375$$

The simplest model for this example would be a line through the origin

$$y_i = \beta x_i + \varepsilon_i \tag{1.5.1}$$

or $\mathbf{y} = \beta\mathbf{x} + \boldsymbol{\varepsilon}$ in vector terms

The normal equation is

$$b \sum x_i^2 = \sum x_i y_i$$

or $(\mathbf{x}^T \mathbf{x})b = \mathbf{x}^T \mathbf{y}$

$$\text{giving} \quad b = (\mathbf{x}^T \mathbf{y})/(\mathbf{x}^T \mathbf{x}) = 0.973 \tag{1.5.2}$$

For each value of x we can calculate the predicted value of y as

$$\hat{\mathbf{y}} = b\mathbf{x} \text{ or } \mathbf{x}b = \begin{vmatrix} 0.5 \\ 1.0 \\ 1.5 \\ 2.0 \\ 2.5 \end{vmatrix} 0.973 = \begin{vmatrix} 0.486 \\ 0.973 \\ 1.459 \\ 1.945 \\ 2.432 \end{vmatrix} \tag{1.5.3}$$

The predicted value can also be written as

$$\hat{\mathbf{y}} = \mathbf{x}b = \mathbf{x}(\mathbf{x}^T\mathbf{x})^{-1}\mathbf{x}^T\mathbf{y} = P\mathbf{y} \tag{1.5.4}$$

The matrix $P = \mathbf{x}(\mathbf{x}^T\mathbf{x})^{-1}\mathbf{x}^T$ is termed the projection matrix. More is said about this in Section 1.7. Notice that for this case with n=5, P is a 5×5 matrix, namely:

$$P = \begin{vmatrix} 0.5 \\ 1.0 \\ 1.5 \\ 2.0 \\ 2.5 \end{vmatrix} (1/13.75)\ (0.5,\ 1.0,\ 1.5,\ 2.0,\ 2.5)$$

$$= \begin{vmatrix} 0.25 & 0.5 & 0.75 & 1.0 & 1.25 \\ 0.5 & 1.0 & 1.5 & 2.0 & 2.50 \\ 0.75 & 1.5 & 2.25 & 3.0 & 3.75 \\ 1.0 & 2.0 & 3.0 & 4.0 & 5.00 \\ 1.25 & 2.5 & 3.75 & 5.0 & 6.25 \end{vmatrix} \ /\ 13.75$$

The vector of residuals is given by

$$\mathbf{e} = (I-P)\,\mathbf{y} = \hat{\mathbf{y}} - \mathbf{y} = \begin{vmatrix} 0.214 \\ 0.177 \\ -0.109 \\ 0.105 \\ -0.132 \end{vmatrix} \qquad (1.5.5)$$

Diagrammatically, we have Figure 1.5.1. Notice that:

(i) $\hat{\mathbf{y}}$ is not affected by the length of $\mathbf{x}$, only its direction. b, however, will be determined by the size of $\mathbf{x}$.
(ii) Given vectors $\mathbf{x}$ and $\mathbf{y}$, $\mathbf{y}$ is uniquely decomposed into two vectors one in the direction of $\mathbf{x}$, that is $\hat{\mathbf{y}}$, and one orthogonal to it, that is $\mathbf{e}$.

The extension to more than one predictor variable is conceptually quite straight forward. The model:

$$y_i = \beta_1 x_{1i} + \beta_2 x_{2i} + \varepsilon_i$$

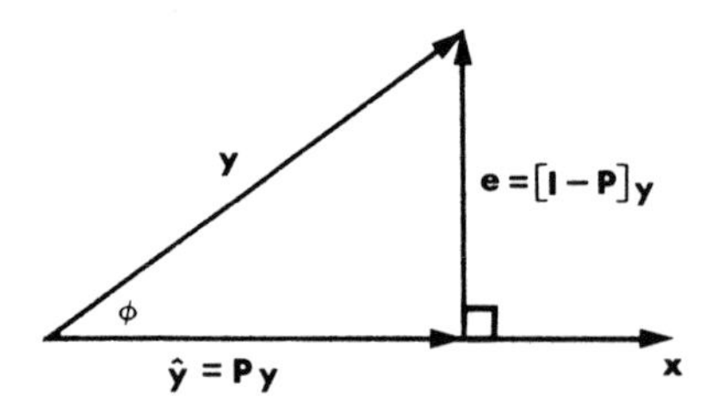

FIGURE 1.5.1 Relationships between observed and predicted y and the residuals, e.

could be written in vector and matrix notations as

$$\mathbf{y} = \beta_1 \mathbf{x}_1 + \beta_2 \mathbf{x}_2 + \varepsilon$$

$$= (\mathbf{x}_1, \mathbf{x}_2) \begin{vmatrix} \beta_1 \\ \beta_2 \end{vmatrix} + \varepsilon$$

$$\text{or } \mathbf{y} = X \boldsymbol{\beta} + \varepsilon \qquad (1.5.6)$$

For n observations, X would be a n×2 matrix. The normal equations, of which there would be two, are

$$X^T X \mathbf{b} = X^T \mathbf{y} \qquad (1.5.7)$$

In full, the normal equations are

$$\mathbf{x}_1^T \mathbf{x}_1 b_1 + \mathbf{x}_1^T \mathbf{x}_2 b_2 = \mathbf{x}_1^T \mathbf{y}$$

$$\mathbf{x}_2^T \mathbf{x}_1 b_1 + \mathbf{x}_2^T \mathbf{x}_2 b_2 = \mathbf{x}_2^T \mathbf{y} \qquad (1.5.8)$$

The analogy with (1.5.3) is

$$\hat{\mathbf{y}} = X (X^T X)^{-1} X^T \mathbf{y} = P \mathbf{y} \qquad (1.5.9)$$

P or P(X) is the projection onto the plane $X = (\mathbf{x}_1, \mathbf{x}_2)$. The residuals are

$$\mathbf{e} = (I - P) \mathbf{y}$$

The vectors can be represented geometrically as in Figure 1.5.2.

If X involves more than two vectors, then X represents a hyperplane. We cannot draw this in two dimensions, but it is convenient to retain the above figure as a picture of the relationships between **y**, **e** and **y**. In particular, we should be clear about the equivalence of the following expressions:

predicted values of **y** = projection of **y** on X

residual = projection of **y** on a direction orthogonal to X

= **y** adjusted for X

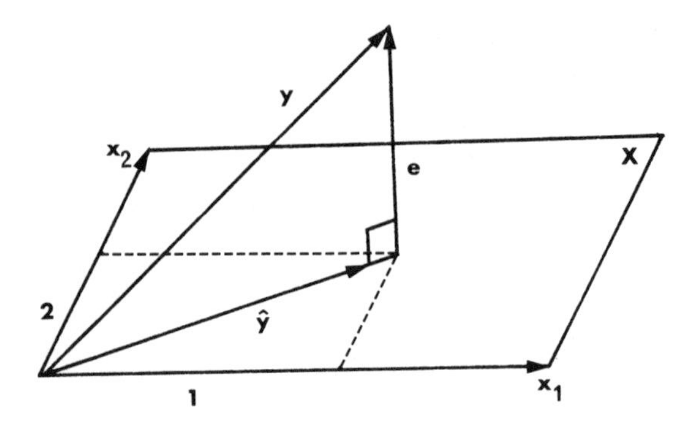

FIGURE 1.5.2 The relationships between observed and predicted y and the residuals, e, in the case of more than 1 predictor variable.

For the special case in which $\mathbf{x}_1$ is orthogonal to $\mathbf{x}_2$, written $\mathbf{x}_1 \perp \mathbf{x}_2$, then $\mathbf{x}_1^T\mathbf{x}_2 = 0$ and the normal equations become

$$\mathbf{x}_1^T \mathbf{x}_1 b_1 = \mathbf{x}_1^T \mathbf{y} \quad \text{and} \quad \mathbf{x}_2^T \mathbf{x}_2 b_2 = \mathbf{x}_2^T \mathbf{y} \qquad (1.5.10)$$

These equations would be the same as those arising from regressing y on x_1 alone and regressing y on x_2 alone. Furthermore,

$$\begin{aligned} P &= X\,(X^TX)^{-1}\,X^T \\ &= (\mathbf{x}_1\ \mathbf{x}_2) \begin{vmatrix} \mathbf{x}_1^T\mathbf{x}_1 & \mathbf{x}_1^T\mathbf{x}_2 \\ \mathbf{x}_2^T\mathbf{x}_1 & \mathbf{x}_2^T\mathbf{x}_2 \end{vmatrix}^{-1} \begin{vmatrix} \mathbf{x}_1^T \\ \mathbf{x}_2^T \end{vmatrix} \\ &= \mathbf{x}_1(\mathbf{x}_1^T\mathbf{x}_1)^{-1}\mathbf{x}_1^T + \mathbf{x}_2(\mathbf{x}_2^T\mathbf{x}_2)^{-1}\mathbf{x}_2^T \quad \text{if} \quad \mathbf{x}_1 \perp \mathbf{x}_2 \\ &= P_1 + P_2 \end{aligned}$$

Then $\hat{\mathbf{y}} = P\,\mathbf{y} = P_1\,\mathbf{y} + P_2\,\mathbf{y}$

These results are clear from Figure 1.5.2. $P_1\mathbf{y}$, represented by (1) is the projection of $\mathbf{y}$ on $\mathbf{x}_1$ and $P_2\mathbf{y}$, represented by (2) is the projection of $\mathbf{y}$ on $\mathbf{x}_2$.

Also $\hat{\mathbf{y}}$ is obtained by adding the vectors (1) and (2). This neat additive property does not follow, of course, if the predictor variables are not orthogonal.

1.6 DEVIATIONS FROM MEANS

It is common practice when fitting a model to use the original (also called raw) data and to include in the model a y-intercept term (also called a constant, or general mean). Most computer programs would convert the raw data to deviations from the mean, as these are used in such statistics as the correlation coefficient. Converting to deviations has the advantage of removing a parameter from the model, making it easier to manipulate. Sometimes an examination of the deviations shows up trends which are not as clearly noticeable in the raw data. Problem 2.1 is an example where deviations from the mean prove useful. It turns out that the estimated coefficients will be the same for the raw data with constant term as with the data in deviation form.

For this section we change our notation slightly to make it clear whether we are referring to the raw data (which we indicate by capital letters, X, Y, etc) or deviations from means (lower case x, y, etc).

1.6.1 Estimates

Ignoring subscripts for simplicity, we can write for the case of one predictor variable,

$$x = X - \bar{X} \quad \text{and} \quad y = Y - \bar{Y}$$

where $\bar{X}$ and $\bar{Y}$ are the sample means. For the model

$$y = \beta x + \varepsilon \tag{1.6.1}$$

we saw in (1.2.7) the least squares estimates are given by

$$\begin{aligned} b &= \sum x_i y_i \,/ \sum x_i^2 && \text{(from 1.2.7)} \\ &= S_{xy} / S_{xx} && \text{(defined by 1.2.13)} \end{aligned} \tag{1.6.2}$$

Notice that the predicted value of y in this case is

$$\hat{y} = b\ x$$

If we change back to the unscaled variables, we have

$$(\hat{Y}-\bar{Y}) = b\ x = b(X-\bar{X}) \tag{1.6.3}$$

Rearranging gives

$$\begin{aligned}\hat{Y} &= \bar{Y} - b\ \bar{X} + b\ X \\ &= a + b\ X\end{aligned} \tag{1.6.4}$$

So we see that the model of (1.6.1) gives rise to the same estimates, b and a, as the model of (1.2.10), namely

$$b = S_{xy}/S_{xx} \quad \text{and} \quad a = \bar{Y} - b\ \bar{X}$$

1.6.2 Vector Interpretation

If the deviations from the sample are written as vectors, that is **x** and **y**, then the cosine of the angle between these two vectors is the familiar correlation coefficient.

$$\cos\theta = (\mathbf{x}^T\mathbf{y})/\surd[(\mathbf{x}^T\mathbf{x})(\mathbf{y}^T\mathbf{y})] = r_{xy} \tag{1.6.5}$$

The vector **y**, the vector of deviations of Y from its mean, can be thought of as the vector of residuals when **Y** is regressed on a constant term, that is on **1**, a vector of ones.

$$\mathbf{Y} = \alpha\mathbf{1} + \varepsilon \tag{1.6.6}$$

$$a = (\mathbf{1}^T\mathbf{1})^{-1}\ \mathbf{1}^T\mathbf{Y} = \bar{Y}, \text{ the sample mean.}$$

The predicted values each equal the sample mean, giving the vectors of predicted values and residuals as

$$\hat{\mathbf{Y}} = a\,\mathbf{1} = \begin{vmatrix} \bar{Y} \\ \bar{Y} \\ \vdots \\ \bar{Y} \end{vmatrix} = \bar{Y}\,\mathbf{1}; \quad \text{residuals} = \mathbf{Y}-\hat{\mathbf{Y}} = \begin{vmatrix} Y_1-\bar{Y} \\ Y_2-\bar{Y} \\ \vdots \\ Y_n-\bar{Y} \end{vmatrix} = \mathbf{y} \tag{1.6.7}$$

We sometimes refer to the deviations, y , as Y adjusted for its mean. Similarly, x can be thought of as X adjusted for its mean.

From Section 6.1, we argue that the same estimates of β can be obtained from the two models:

(i) Include a constant term: $Y = \alpha + \beta X + \varepsilon$
(ii) Adjust each variable for its mean (or for the vector 1):
$y = \beta x + \varepsilon$

This statement could obviously extend to any number of predictor variables. It is of more interest to notice that there is nothing particularly special about a vector of 1's, so that the same argument should apply if either a vector **x** is included in the model or if all the other variables are adjusted for it.

By adjusted, we mean that each variable is regressed on x and the residuals obtained. We shall return to this point in Chapter 3 when we consider entering variables in the model sequentiallly.

1.6.3 Normal Equations

If all variables are expressed as deviations from the means, the normal equations of (1.2.14) become

$$\begin{aligned} b_1S_{11} + b_2S_{12} + \cdots + b_kS_{1k} &= S_{y1} \\ b_1S_{12} + b_2S_{22} + \cdots + b_kS_{2k} &= S_{y2} \\ \vdots \qquad \vdots \qquad\qquad \vdots \quad &\quad \vdots \\ b_1S_{1k} + b_2S_{2k} + \cdots + b_kS_{kk} &= S_{yk} \end{aligned} \tag{1.6.8}$$

A convenient way to write the general linear model is

$$\mathbf{y} = \mathbf{1}\,\beta_0 + X\,\boldsymbol{\beta} + \boldsymbol{\varepsilon} \tag{1.6.9}$$

Here, β_0 is a constant, or intercept term, and X is expressed as deviations from sample means. Alternatively, we could write the dependent variable in terms of deviations to give

$$\mathbf{y} = X\boldsymbol{\beta} + \boldsymbol{\varepsilon} \qquad (1.6.10)$$

1.7 AN EXAMPLE - VALUE OF A POSTAGE STAMP OVER TIME

The value of an Australian stamp (1963 £2 sepia colored) given, in £ sterling, by the Stanley Gibbons Catalogues is shown in Table 1.7.1 for the years 1972-1980. The aim is to fit a model to the value of the stamp over time.

Let the time be x = 0 for 1972, x = 1 for 1973, etc. In this example, interest probably centres on the relative value of the stamp from one year to the next. Alternatively, if there is interest in the investment value of the stamp over the period 1972-1980, it may be useful to express the value as y, the value relative to that of 1972. The values of y and x are also shown above and are graphed in Figure 1.7.1. The relationship between y and x is not a linear one. When y is transformed by taking natural logarithms, a strong linear trend is apparent as shown by Figure 1.7.2. The predicted values of ln y and the residuals are shown in Table 1.7.2. The prediction equation is

$$\ln y = -0.222 + 0.342\ x + e \qquad (1.7.1)$$

TABLE 1.7.1. Value of Australian Stamp

Year	Value £ Sterling	x	y
1972	10	0	1.0
1973	12	1	1.2
1974	12	2	1.2
1975	22	3	2.2
1976	25	4	2.5
1977	45	5	4.5
1978	75	6	7.5
1979	95	7	9.5
1980	120	8	12.0

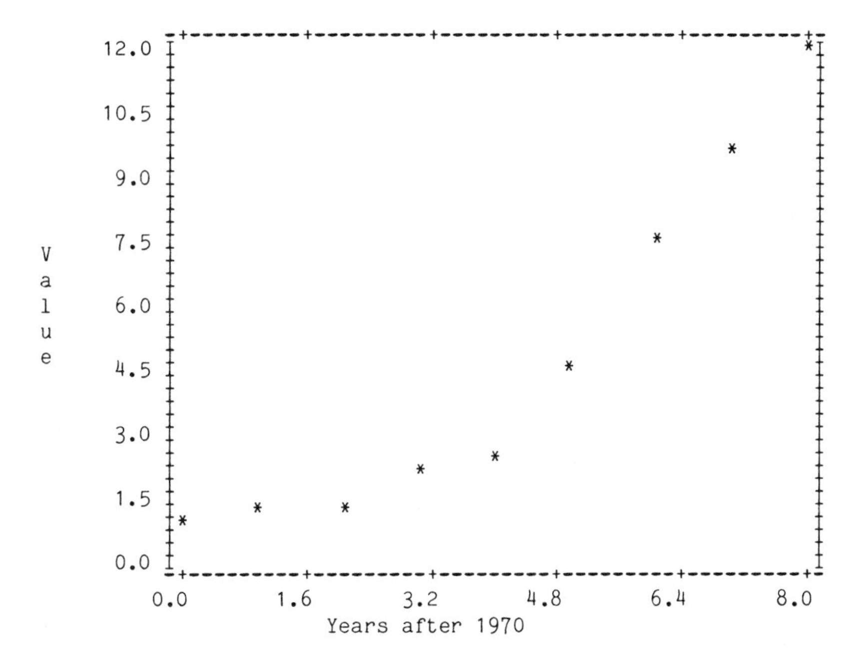

FIGURE 1.7.1 The relative value of the stamp against time.

Of course, the constant term could be omitted and the graph forced to pass through the origin.

The residuals fall into a reasonable horizontal band so that there is little evidence to contradict the assumptions that the deviations are distributed with mean zero and constant variance. However, there is a marked pattern in the residuals for their signs are:

+ + - - - + + + -

Clearly, the residuals are positively correlated (as a positive residual is often followed by a positive residual, and a negative residual followed by a negative). With the small number of observations in the sample, we should hesitate to make dogmatic statements about the model but the pattern in the residuals may suggest that

(i) The population mean is not correct.
The residuals could indicate that a cubed term would remove

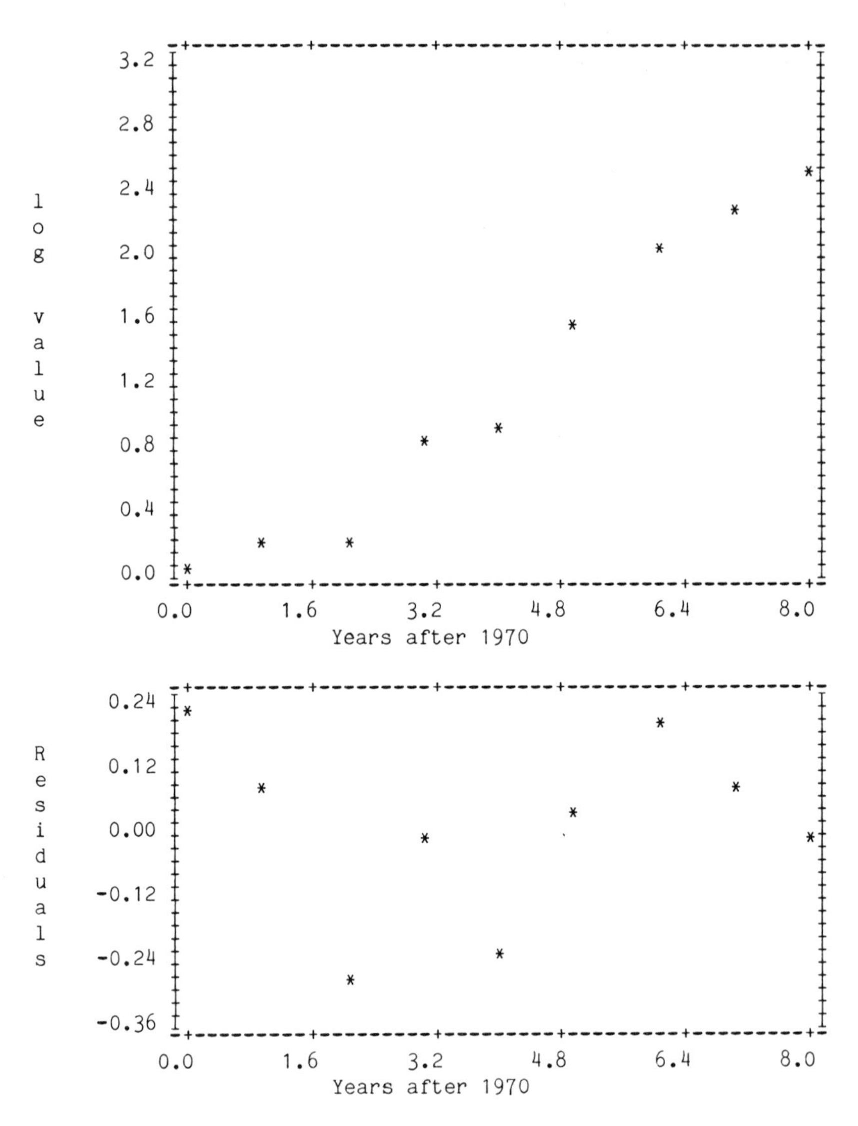

FIGURE 1.7.2 Log (base e) of stamp value, y, and residuals against time, x.

TABLE 1.7.2 Predicted Values and Residuals : Transformed Data

x	ln y	Predicted ln y	Residuals
0	0.000	-0.222	0.222
1	0.182	0.121	0.061
2	0.182	0.463	-0.281
3	0.788	0.805	-0.017
4	0.916	1.147	-0.231
5	1.504	1.489	0.015
6	2.015	1.831	0.184
7	2.251	2.173	0.078
8	2.485	2.516	-0.031

some of the cyclical variation. The prediction equation is now

$$\ln y = -0.149 + 0.296\,x + 0.001\,x^3 + e \qquad (1.7.2)$$

Notice that the estimated coefficients for the constant and linear terms have been changed somewhat by the introduction of the cubed term. The estimated coefficient for the cubed term seems insignificant but remember that x cubed can take large values, e.g., 343 when x = 7. With this model the signs of the residuals are:

+ + − + − + + − −

There is still a pattern in these residuals but the more fluctuating signs reduce the possibility that the deviations are related and not independent.

(ii) The population mean is correct, but deviations are not independent. This may not worry us unduly as under reasonable assumptions the trend line remains a satisfactory one. To be precise, the predicted values under our false assumption of independence of deviations remains unbiased, but the variance may be high as we have not taken into account the additional fact of related deviations. What should be done about this possible dependence of the deviations? Some approaches could be:

(a) Accept the trend line but take into account the dependence. From (1.7.1) when x = 9, we would estimate ln y as 2.856. From the pattern of residuals we might guess, however, that the residual will again be negative for x = 9.

(b) In some situations it may be sensible to attempt a smoothing process such as a running average of 3 points. This example would have pairs of observations

$$(x, \text{average } \ln y) = (1,[0 + 0.182 + 0.182]/3),\quad (2,[0.182 + 0.182 + 0.788]/3) \text{ etc.}$$

End points could be treated slightly differently or the number of observations reduced from 9 to 7. This gives the trend line

$$\text{average } \ln y = -0.351 + 0.368\, x + e \qquad (1.7.3)$$

Unfortunately, in this example with positively correlated residuals this procedure exacerbates the problem as the signs of the residuals are

+ - - - - + +

(c) Reference will be made later to this general problem of autocorrelation of deviations, as it is usually called.

PROBLEMS

1.1 Fit the model of (1.7.2) to check that the pattern of residuals is as stated.

1.2 For the data of Example 1.7, suggest another possible transformation of the dependent variable and estimate coefficients.

1.3 A researcher fitted a simple model and obtained the following graph of residuals against predicted values.
Some texts suggest that this could only arise by a mistake. No mistake was made in calculation, however, so what model had been fitted?

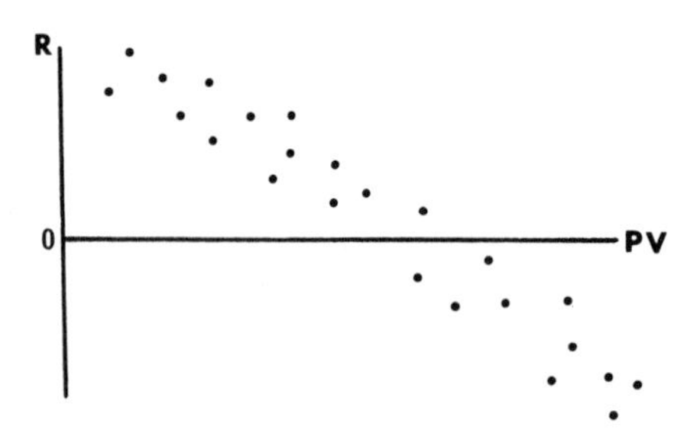

1.4 The following pairs of data (x,y) refer to a single calf with x the number of days after birth and y the serum gamma glutamyl transpeptidase (GGT) in Iu/l at various times after birth.

x	1	4	6	8	11	13	15	18	21	25	28	32
y	563	236	167	153	104	85	71	50	43	28	21	19

(i) Plot y against x.
(ii) Suggest a transformation of y.
(iii) Find the least squares fit for your model above. Plot residuals against predicted values of y.
(iv) Does your model seem to fit the data? Are there some aspects of the model fitting with which you are not happy?

1.5 The following observations are the results students gained in three math papers. They take 202 and 203 in their second year and then 303 in their third year.

202	203	303
73	85	85
59	59	43
68	64	74
57	54	41
44	60	50
61	57	52
84	55	50
63	52	68
59	56	50
69	64	82
93	91	90
56	52	55
72	58	67
73	72	78

To help decide whether the results in the second year papers are good predictors of the outcome in 303, we could fit the model

$$y = \beta_0 + \beta_1 x_1 + \beta_2 x_2 + \varepsilon$$

Here, y is the result in 303 and x_1 and x_2 are in deviation form corresponding to the results in 202 and 203.

(i) Calculate the estimates b_0, b_1, and b_2.
(ii) Find the predicted outcome, $\hat{y}$, for each student in 303. Does any value appear unusual?
(iii) Which pairs of the four vectors $\mathbf{1}$, $\mathbf{x}_1$, $\mathbf{x}_2$, $\mathbf{y}$ are orthogonal?

Could the projection matrix, P, onto the three predictor variables be written as the sum of individual projection matrices as in Section 1.5?

2

GOODNESS OF FIT OF THE MODEL

2.1 INTRODUCTION

In Chapter 1, we considered fitting models to data. To answer the question of how well a model fits the data, we need to develop statistical tests and, to do so, we lean heavily on the assumption that the deviations have independent normal distributions. For our purposes, the main benefit of the normality assumption is that we can find the distribution of estimates and perform test of significance. The theorems listed in Appendix B will be of assistance in this.

For the general linear model

$$\mathbf{y} = \mathbf{1}\alpha + X\boldsymbol{\beta} + \boldsymbol{\varepsilon} \tag{2.1.1}$$

where α is a constant term, X is an $n{\times}k$ matrix of known constants, deviations from their mean, and $\boldsymbol{\varepsilon}$ are the deviations, assumed to be independent and normally distributed with mean zero and constant variance of σ^2, that is $\boldsymbol{\varepsilon} \sim N(0,\sigma^2 I)$. I is an $n{\times}n$ identity matrix with all diagonal elements equal to 1 and off diagonal elements equal to zero. Thus, the covariances are zero and the variances are all equal.

With these assumptions, we have, in effect, made assumptions about **y**. The mean, or expected value, of **y** is $X\boldsymbol{\beta}$ as $E(\boldsymbol{\varepsilon})$ is 0. The variance of **y** is the variance of $\boldsymbol{\varepsilon}$, $\sigma^2 I$.

2.2 COEFFICIENT ESTIMATES FOR UNIVARIATE REGRESSION

Consider the simple univariate regression with a constant term included, that is

$$\mathbf{y} = \alpha\mathbf{1} + \beta\mathbf{x} + \boldsymbol{\varepsilon} \quad \text{with} \quad \boldsymbol{\varepsilon} \sim N(0, \sigma^2 I) \quad \text{or} \quad \mathbf{y} \sim N(\alpha\mathbf{1} + \beta\mathbf{x}, \sigma^2 I) \qquad (2.2.1)$$

The least squares estimate of β is

$$b = (\mathbf{x}^T\mathbf{x})^{-1}\mathbf{x}^T\mathbf{y}$$

Note that since the x's are deviations from their means

$$b = S_{xy} / S_{xx} \qquad \text{(from 1.6.2)}$$

and $S_{xy} = \sum x_i (y_i - y) = \sum x_i y_i = \mathbf{x}^T\mathbf{y}$

As b is a linear combination of the observed y values, then Appendix B 1 can be invoked to give

$$E(b) = S_{xx}^{-1}\mathbf{x}^T(\alpha\mathbf{1} + \beta\mathbf{x}) = \beta \quad \text{as} \quad \mathbf{x}^T\mathbf{1} = 0$$

$$\text{var}(b) = S_{xx}^{-1}(\mathbf{x}^T\mathbf{x})S_{xx}^{-1}\sigma^2 = S_{xx}^{-1}\sigma^2$$

σ^2 can be estimated by the sample variance of the residuals, namely

$$s^2 = \sum e_i^2 / (n-2) \qquad (2.2.2)$$

This has n-2 degrees of freedom as there are n observations, but two parameters, α and β, to be estimated.

These results can be combined to give statistics for inferences about β. The standard error of b, s_b, is estimated by $s/\sqrt{S_{xx}}$. Therefore

$$(b - \beta)/s_b \sim t_{n-2}, \quad \text{the 'Student' t distribution} \tag{2.2.3}$$

so that confidence intervals for β are given by

$$b \pm t_{n-2} s_b \tag{2.2.4}$$

and the hypothesis, H: $\beta = 0$, can be tested by t:

$$t_{n-2} = b/s_b \tag{2.2.5}$$

Instead of including the constant term in the model above, we could have adjusted both variables by expressing them as deviations from their means.

2.3 COEFFICIENT ESTIMATES FOR MULTIVARIATE REGRESSION

For the model

$$\begin{aligned} \mathbf{y}_0 &= \beta \; \mathbf{1} + \beta_1 \mathbf{x}_1 + \beta_2 \mathbf{x}_2 + \cdots + \beta_k \mathbf{x}_k + \boldsymbol{\varepsilon} \\ &= X \beta + \varepsilon \quad \text{with} \quad \varepsilon \sim N(0, \sigma^2 I) \end{aligned} \tag{2.3.1}$$

the estimated coefficient vector is

$$\mathbf{b} = (X^T X)^{-1} X^T \mathbf{y}$$

Each coefficient in **b** is a linear combination of the y values so that, by Property 2, Appendix B2, it is distributed normally. In fact,

$$\mathbf{b} \sim N(\boldsymbol{\beta}, (X^T X)^{-1} \sigma^2) \tag{2.3.2}$$

For convenience, we write $(X^T X)^{-1}$ as the matrix $C = \{c_{ij}\}$. A particular coefficient estimate,

$$b_i \sim N(\beta_i, c_{ii}\sigma^2) \tag{2.3.3}$$

To test a particular coefficient or determine confidence intervals, we use the method of Section 2.2. We may be interested in a linear combination of these estimates. This is also a linear combination of the y values or their deviations. For such a linear combination we have, again from Appendix B 2,

$$\mathbf{a}^T\mathbf{b} \sim N(\mathbf{a}^T\boldsymbol{\beta}, \mathbf{a}^T C \mathbf{a}\, \sigma^2)$$

For example, we may wish to test:

$$H: \beta_1 = \beta_3, \quad \text{or} \quad H: \beta_1 - \beta_3 = 0$$

in the model

$$\mathbf{y} = \beta_0 + \beta_1 \mathbf{x}_1 + \cdots + \beta_4 \mathbf{x}_4 + \boldsymbol{\varepsilon}$$

Then $\mathbf{a}^T = (0, 1, 0, -1, 0)$ and $\mathbf{a}^T C\mathbf{a} = c_{11} - 2c_{13} + c_{33}$. This is not surprising as

$$\begin{aligned} \text{var}(b_1 - b_3) &= \text{var } b_1 + \text{var } b_3 - 2\text{cov }(b_1, b_3) \\ &= (c_{11} + c_{33} - 2c_{13})\, \sigma^2 \end{aligned}$$

To test the hypothesis $H: \beta_1 = \beta_3 = 0$ we cannot follow the above approach as here there are really two hypotheses,

$$H_1: \beta_1 = 0 \quad \text{and} \quad H_2: \beta_3 = 0$$

We could test either of these separately but their tests will not be independent. To test these joint hypotheses, we could use the method of reduced models of Section 2.9.

2.4 ANOVA TABLES

The analysis of variance table, ANOVA, represents the components of the variation of the dependent variable, y. In Section 1.5, we showed that, for any number of predictor variables, we can divide the vector of observations, y, into two orthogonal parts consisting of the predicted values and the residuals, which we can represent by Figure 1.5.1.

As the vectors $\mathbf{e}$ and $\hat{\mathbf{y}}$ are orthogonal, $\mathbf{e} \perp \hat{\mathbf{y}}$, then from Pythagoras' Theorem we know that

$$(\text{length } \mathbf{y})^2 = (\text{length } \hat{\mathbf{y}})^2 + (\text{length } \mathbf{e})^2 \quad \text{or}$$

$$\mathbf{y}^T\mathbf{y} = \hat{\mathbf{y}}^T\hat{\mathbf{y}} + \mathbf{e}^T\mathbf{e} \tag{2.4.1}$$

If there are k predictor terms in the model, but no constant (intercept) term, then the sums of squares and degrees of freedom, d.f., can be displayed as in Table 2.4.1.

Almost invariably it is deviations from the mean which the model is required to explain, so a constant term is included in the model. We have seen from Section 1.6.2, this has the effect of adjusting each variable for its mean, and the sum of squares for the mean is given by

$$\text{SS(Mean)} = \mathbf{y}^T\mathbf{1}\,(\mathbf{1}^T\mathbf{1})^{-1}\,\mathbf{1}^T\mathbf{y} = (\Sigma\, y_i)^2/n = n\,\bar{y}^2$$

If the first variable is a constant then subtracting SS(Mean) from SSR and SST gives the sums of squares and mean squares adjusted for the mean. The same ANOVA would result from the regression of y, in deviation form, on k predictor variables, also in deviation form as shown in Table 2.4.2. We shall usually assume that the variables are in deviation form as this facilitates the discussion, but in

TABLE 2.4.1 ANOVA With NO Constant Term (y Values Are Raw Scores)

Source of Variation	d.f.	Sums of Squares	Mean Squares
Regression	k	$\hat{\mathbf{y}}^T\hat{\mathbf{y}} = \mathbf{y}^T P\,\mathbf{y}$	SSR/k
Error	n-k	$\mathbf{e}^T\mathbf{e} = \mathbf{y}^T(I-P)\mathbf{y}$	SSE/(n-k)
Total	n	$\mathbf{y}^T\mathbf{y}$	SST/n

TABLE 2.4.2 ANOVA With a Constant Term (y Values Are Deviations From the Mean)

Source of Variation	d.f.	Sums of Squares	Mean Squares
Regression	k	$\hat{\mathbf{y}}^T\hat{\mathbf{y}} = \mathbf{y}^T P\, \mathbf{y}$	SSR/k
Error	n-k-1	$\mathbf{e}^T\mathbf{e} = \mathbf{y}^T(I-P)\mathbf{y}$	SSE/(n-k-1)
Total	n-1	$y^T y$	SST/(n-1)

practice we could merely include an overall mean in the model and proceed with the original values of the variables.

The three sums of squares again represent the squares on the sides of the triangle in Figure 1.5.1. We have used "Error" rather than "Residual". This avoids some confusion in termination as we use

SSR to mean sum of squares for regression

SSE to mean sum of squares for error (or residuals)

SST to mean sum of squares for total

One immediate use of the ANOVA table is that it provides the mean square for error, MSE = SSE/(n-k-1). This is an unbiased estimate of the variance, which is the variance of each deviation, or the variance of y about the regression line.

2.5 THE F-TEST

The ANOVA provides an F-statistic to test whether the model fits the data. The F-statistic is basically a comparison of two estimates of variance. The MSE provides an estimate of the variance which is unbiased regardless of the coefficient vector . The mean square for regression, MSR, provides another estimate of this variance but it is only unbiased if the coefficient vector is zero.

The model with a constant term can be written as (2.1.1) or, with all variables in deviation form, as

$$\mathbf{y} = X\,\boldsymbol{\beta} + \boldsymbol{\varepsilon} \tag{2.5.1}$$

Notice that

$$E(\hat{\mathbf{y}}) = E(P\mathbf{y}) = X\,(X^TX)^{-1}\,X^TX\,\boldsymbol{\beta} = X\,\boldsymbol{\beta}$$

and this will equal zero if $\boldsymbol{\beta} = 0$. Thus, to test H: $\boldsymbol{\beta} = 0$ we use F = MSR/MSE. If, in fact, $\boldsymbol{\beta} \neq 0$ SSR is inflated and F tends to be large.

If the F value is small to the point that H cannot be rejected, we would have serious doubts about our model. If the F value is large, we are at least assured that the model is feasible. Of course, it does not mean that this particular model is in any way optimal. It could be, for example, that one variable is providing the bulk of the sums of squares for regression while other variables in the model are quite irrelevant.

2.6 THE COEFFICIENT OF DETERMINATION

With the model in deviation form, we define the coefficient of determination as:

$$R^2 = SSR\ /\ SST = \sum (\hat{y}_i - \bar{y})^2 / \sum (y_i - \bar{y})^2 \qquad (2.6.1)$$

This coefficient indicates the proportion of the variation of $\mathbf{y}$ explained by X. Another way of looking at it is in terms of the angle between the vectors of predicted and observed y. If the vectors in Figure 1.5.1 were expressed as deviations from their means we see that

$$\cos\theta \text{ is the ratio } (\text{length } \hat{\mathbf{y}})/(\text{length } \mathbf{y}) = R$$

This value, R, has the form of a correlation coefficient between $\hat{\mathbf{y}}$ and $\mathbf{y}$ and is referred to as the multiple correlation coefficient. The model is obviously a good fit if the cosine is close to zero, which means that R^2 would be close to one.

As a rule of thumb, the coefficient of determination should be greater than 0.5 for us to have much confidence in the model. After all, if it equals 0.5 then half of the variation of y is not explained by the model. Its size depends on the sample size and the number of parameters to be estimated. For this reason, some people like to work with an adjusted value and a rationale for this follows.

2.6.1 Adjustment for Degrees of Freedom

The coefficient R^2 measures the proportion of the variation in y which is explained by the predictor variables. Actually, it overestimates this proportion and the adjustment suggested here aims to correct this.

If the y values were entirely random in the space of n dimensions (the deviations from their mean would then be in an n-1 dimensioned sub-space) the k predictor variables would still explain some variation in y, on average $k/(n-1)$. R^2 is corrected for this random effect by subtracting $k/(n-1)$. This is then scaled to give a value of 1 (perfect explanation of y) when $R^2 = 1$. Finally

$$\text{adj } R^2 = [R^2 - k/(n-1)][n-1]/[n-k-1] \qquad (2.6.2)$$

This adjusted value could even be negative if R^2 is small, which highlights the only problem with it. What would a negative value mean? On the other hand the unadjusted value has a clear interpretation as the proportion of the variance of y explained by the predictor variable.

2.7 PREDICTED VALUES OF Y AND CONFIDENCE INTERVALS

2.7.1 The Univariate Case

Consider again the simple model

$$\mathbf{y} = \alpha \mathbf{1} + \beta \mathbf{x} + \boldsymbol{\varepsilon} \qquad (2.7.1)$$

with $\mathbf{x}$ in deviation form and $\boldsymbol{\varepsilon} \sim N(0, \sigma^2 I)$. Notice that $\mathbf{1}$ is orthogonal to $\mathbf{x}$ so that the least squares estimates of α and β are the same as would be obtained by regressing y separately on $\mathbf{1}$ and $\mathbf{x}$.

$$a = \Sigma y_i/n \quad , \quad b = \Sigma x_i y_i / \Sigma x_i^2 \tag{2.7.2}$$

For a given value of $x = x_0$, the predicted value of y is

$$\hat{y}_0 = a + b x_0 \tag{2.7.3}$$

A number of points should be kept in mind relating to this expression.

(i) The x_0 need not be one of the x in the sample, but there are obvious dangers in predicting outside of the range of the sample. For one thing, the relationships between x and y may be linear for the x values of the sample but the relationship may change outside of these values, as in Figure 2.7.1. One example of this could be the cancer causing effects of low doses of radiation. The incidence of cancer in this situation would be so low that it would be difficult to measure, requiring a very large sample size. To facilitate the research one could work at higher dosage rates (x) and hope that one could extrapolate down to lower dosage rates, but this procedure is fraught with danger.

(ii) The predicted value of y depends on <u>all</u> the y values in the sample. We shall comment further on this in Chapter 4 where we shall discuss <u>sensitive</u> or <u>high leverage</u> values of x, by which we mean those x's which have a very large effect on the predicted values of y.

(iii) $\hat{y}$ estimates the <u>mean</u> value of y when $x = x_0$.

(iv) y is a linear combination of the y values in the sample.

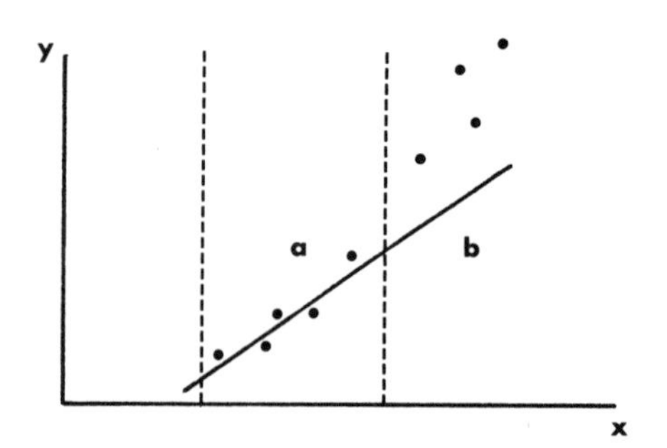

FIGURE 2.7.1 Prediction outside of the range of the sample.
a. In this region a straight line fits the data well.
b. If the line is extrapolated to this region which is outside the range of the sample, then predictions would clearly be inaccurate.

(v) In this case the covariance beween a and b is zero as **1** and **x** are orthogonal. Thus

$$\text{var } \hat{y} = \text{var } \hat{a} + x^2 \text{ var } \hat{b}$$

and

$$\text{var } \hat{y}_0 = \sigma^2/n + x_0^2\sigma^2/\Sigma\, x_i^2 = \sigma^2(1/n + x_0^2/\Sigma\, x_i^2) \tag{2.7.4}$$

The standard error of $\hat{y}$ is the square root of the expression in (2.7.4) with the variance estimated by MSE. We will write this as s*. A confidence interval for the mean value of y when $x = x_0$, is

$$\hat{y}_0 \pm t\, s^* \tag{2.7.5}$$

Here t is the appropriate percentile of the t statistic with the same number of degrees of freedom as there are in MSE. The width of the confidence interval is 2ts*.

We illustrate this with the simple calibration data of Example 1.3.1. We have expressed the predictor values as deviations from the mean. We also refer to a future y. We will consider this in Section 2.7.3. We also need the quantities

$$\Sigma x_i^2 = 2.5 \qquad s^2 = 0.0153 \qquad t_{.975}(3) = 3.18$$

Notice that the confidence intervals are narrowest when the x values as deviations are smallest, which in this case is the central values of x.

2.7.2 The Multivariate Case

With more than one predictor variable we can call on the model in (2.1.1), namely

$$\mathbf{y} = \mathbf{1}\alpha + X\boldsymbol{\beta} + \boldsymbol{\varepsilon} \tag{2.7.6}$$

with the predictor variables expressed as deviations from their means $\mathbf{1}^T\mathbf{X} = 0$, giving the estimated coefficients

$$a = \bar{y} \qquad \text{with var } a = \sigma^2/n$$

$$\mathbf{b} = (X^T X)^{-1} X^T \mathbf{y} \qquad \text{with var } b = (X^T X)^{-1} \sigma^2$$

The predicted value of y for a particular value of x_0, say

$$\mathbf{x}_0^T = (1, x_{0_1}, x_{0_2}, \ldots, x_{0_k}) \text{ is}$$

$$\hat{y}_0 = a + b_1 x_{0_1} + b_2 x_{0_2} + \cdots + b_k x_{0_k} = a + \mathbf{x}_0 \mathbf{b}^T \qquad (2.7.7)$$

$$\text{var } \hat{y}_0 = \text{var } a + \mathbf{x}_0^T \text{var } \mathbf{b} \, \mathbf{x}_0$$

$$= \sigma^2/n + \mathbf{x}_0^T (X^T X)^{-1} \mathbf{x}_0 \sigma^2 \qquad (2.7.8)$$

Confidence intervals for the mean value of y will follow in exactly the same way as in the univariate case. Again, the width of the confidence interval near the centre of the distribution will be narrower than at the extreme points of x. This is related to the topic of sensitive, or high leverage, points which we will take up in Chapter 4.

It is often convenient to write the dependent variable in deviation form giving the model

$$\mathbf{y} = X \boldsymbol{\beta} + \boldsymbol{\varepsilon}$$

with the vector of predicted values

$$\hat{\mathbf{y}} = X \mathbf{b}$$

and the variance-covariance matrix as

$$\text{var } \hat{\mathbf{y}} = X (X^T X)^{-1} X^T \sigma^2 \qquad (2.7.9)$$

2.7.3 Individual, or Future, Values

So far we have stressed that a predicted value estimates the mean of y at a given value of x. It may be of some interest to estimate an individual value of y in the future. The best estimate of an individual value of y will be the same as the estimate of the mean value of y. A difference will arise in the variance, for we know that an individual value of y varies about the regression line with a variance of σ^2.

We could write

$$\hat{y}_f = \hat{y} + e_f \tag{2.7.10}$$

giving

$$\text{var } \hat{y}_f = \text{var } \hat{y} + \text{var } e_f$$

The Equation (2.7.4) would now become, for the univariate case

$$\text{var } \hat{y}_f = \left(1 + 1/n + x_f^2 / \Sigma\, x^2\right) \sigma^2 \tag{2.7.11}$$

This expression leads to wider confidence intervals than in Section 2.7.2.

2.8 RESIDUALS

If the fitted model is the correct one (or close to the correct one), we would expect the residuals to reflect the properties of the deviations. In this chapter, we are mainly concerned with checking the residuals to be assured that the model with its assumptions is reasonable for the data. Is the distribution of the residuals consonant with the assumed distribution of the deviations? Does it appear that the variance about the line is constant? Does it appear that the deviations are independent?

If we recall that the prediction equation is $\mathbf{y} = \hat{\mathbf{y}} + \mathbf{e}$ then, in a sense, the residual and the predicted $\mathbf{y}$ value are opposite faces of the same coin for when one is large, the other is small. We shall not have much to say at this point on the sizes of individual residuals. Large values may indicate that the points are outliers but we shall say more on this in Chapter 4.

2.8.1 The Distribution Of Residuals

For the model $\mathbf{y} = X\,\boldsymbol{\beta} + \boldsymbol{\varepsilon}$, the vector of residuals is $\mathbf{e} = (I-P)\,\mathbf{y}$ which is a linear combination of the observed values. Thus if $\boldsymbol{\varepsilon} \sim N(0, \sigma^2 I)$

(i) $\mathbf{e}$ is distributed normally.
(ii) $E(\mathbf{e}) = (I-P)Xy = 0$ as $PX = X$
This is in line with the assumption that $E(\boldsymbol{\varepsilon}) = 0$.
(iii) $\text{var}\ (\mathbf{e}) = (I-P)\ \text{var}\ \mathbf{y}\ (I-P) = (I-P)\ \sigma^2$
as $I-P$ is idempotent. This contrasts with the assumption that $\text{var}\ \boldsymbol{\varepsilon} = \sigma^2 I$ for the residuals are correlated even though the deviations are not.

Now we will look at the distribution of the individual residuals. We will write

$$P = \{p_{ij}\}$$

$$\text{var}(e_i) = (1-p_{ii})\ \sigma^2 \quad \text{whereas} \quad \text{var}(\varepsilon_i) = \sigma^2$$

$$\text{cov}(e_i, e_j) = -p_{ij}\ \sigma^2 \quad \text{whereas} \quad \text{cov}(\varepsilon_i, \varepsilon_j) = 0$$

As $\mathbf{y} = \hat{\mathbf{y}} + \mathbf{e}$ and $\hat{\mathbf{y}}$ is orthogonal to $\mathbf{e}$ then

$$\text{var}\ y_i = \text{var}\ \hat{y}_i + \text{var}\ e_i \qquad (2.8.1)$$

As we assume $\text{var}\ y_i$ is constant for all i, and as $\text{var}\ \hat{y}_i$ is smallest near the central value of the x's then $\text{var}\ e_i$ will be greatest here. As $(e_i - 0)/\sqrt{((1-p_{ii})s^2)}$ has a t distribution with n-k-1 degrees of freedom we would "expect" with 95% confidence e_i to fall in the interval

$$\pm t_{.975}(n-k-1)s\ \sqrt{(1-p_{ii})} \qquad (2.8.2)$$

This statement is not wholly accurate for while a similar statement could be made about the value of t falling in $\pm t_{.975}$, a large value of e_i would tend to increase the value of s^2. Notice that the confidence intervals are widest for the middle values of x (when p is smallest) giving a 'ballooning effect'. This is the other side of the coin to the confidence intervals for the predicted values of y, which are narrowest at the middle.

The scaled residuals

$$t_i = e_i / s\sqrt{(1 - p_{ii})} \qquad (2.8.3)$$

are called studentized residuals as they approximately follow a student t distribution. Rather than plotting the e's, it would be better to plot t's against y's or x's as the t's should fall within a horizontal band if the model fits the data. However for large data sets the $(1-p_{ii})$ may all be reasonably close to 1 so that a plot of $t_i^* = e_i/s$, called standardized residuals, against $\hat{y}_i$ or x_i will closely approximate a plot of the t's.

If a studentized residual fell outside of its confidence band, it would be tempting to declare that point an outlier or a weird value in relation to the fitted model. Certainly, that point should be carefully scrutinized to ensure that no coding error or such-like has occurred. We should not too hastily remove a point from the model though just because it appears to be an outlier. Its seemingly strange behaviour may be of interest to the researcher. Furthermore, as we shall see in Chapter 4, high leverage points may be present which may be unduly influencing the model.

2.8.2 Dependence of the Deviations: Autocorrelation Function.

We have seen above that, even if the deviations are independent, the residuals will be correlated. If the sample size is reasonably large while k, the number of parameters to be estimated, is small then we would not expect the residuals to be highly correlated. If they are, then we would suspect the independence assumptions of the deviations.

To gain an insight into the relationships between residuals, we can estimate the autocorrelation function (ACF). The first correlation we calculate is between each residual and the residual next to it. We can illustrate this with the five residuals of Exercise 1.3.1. We move the series of residuals one place to the right and we say that it is lagged by one position.

We have the residuals

0.01	0.05	-0.16	0.13	-0.02	
and	0.01	0.05	-0.16	0.13	-0.02

which is the same series lagged by one. If we ignore the 0.01 of the first series and the -0.02 of the second, we can find the correlation between the remaining four pairs as

$$r_1 = \text{first lagged correlation} = -0.68$$

Likewise the series

0.01	0.05	-0.16	0.13	-0.02		
		0.01	0.05	-0.16	0.13	-0.02

gives

$$r_2 = \text{second lagged correlation} = 0.20$$

Note that

(i) Instead of ignoring the endpoints, we could use appropriate correction factors but we shall not worry about these.

(ii) If the deviations were known, instead of being estimated by the residuals from the sample, then their autocorrelations could be found which could be denoted by ρ_1, ρ_2, etc.

(iii) In the simple example above, the sample size is so small that high correlations of the residuals could not be used to infer high correlations among the deviations particularly as Section 2.8.1 indicated that there will always be some correlation among the residuals.

(iv) For large sample sizes, high estimated correlations in the sample would suggest that the deviations of the population are correlated.

The autocorrelation function, ACF, gives these autocorrelations of the residuals. We shall only consider this as a descriptive tool, although tests could be carried out on these autocorrelations. We mention, in passing, the Durbin-Watson test which is based on the first lagged correlation. Their statistic lies between 0 and 4 with a mean of 2, with positive correlations giving values less than 2 and negative greater than 2.

The presence of high autocorrelations does not mean that estimated coefficients are biased but it could suggest that the model could be improved by a transformation or by including other variables which have a cyclical effect on y. Another approach which we shall consider in Chapter 4 is the weighted least squares method.

2.9 REDUCED MODELS

In Section 2.3, we showed that tests on single coefficients could be carried out using the fact that the estimates are normally distributed. We could test a whole series of hypotheses about a model in this way but unless we are very careful these tests would not be independent so that we could not be sure that the significance level was as stated.

A more general approach is to write the normal equations for the full model and from these find the sum of squares for residuals, SSE. Under the null hypothesis, the model becomes restricted to the reduced model and from the resulting normal equations we find SSE (Reduced). We test whether the sum of squares for residuals has been reduced significantly by forming the F-statistic

$$F_{r,n} = [(SSE(Reduced) - SSE)/r]/[SSE/n] \tag{2.9.1}$$

Here n is the degree of freedom associated with SSE and r is the number of restrictions placed on the model by the null hypothesis.

Consider a model with three predictor variables expressed as deviations from their means,

$$\mathbf{y} = X\boldsymbol{\beta} = \beta_1\mathbf{x}_1 + \beta_2\mathbf{x}_2 + \beta_3\mathbf{x}_3 + \boldsymbol{\varepsilon} \tag{2.9.2}$$

Notice the following structure of the normal equations where R1 refers to row 1 and involves x_1, C2 refers to column 2 and involves x_2, and so on.

	C1		C2		C3		C4
R1	$\Sigma x_1^2 b_1$	+	$\Sigma x_1 x_2 b_2$	+	$\Sigma x_1 x_3 b_3$	=	$\Sigma x_1 y$
R2	$\Sigma x_1 x_2 b_1$	+	$\Sigma x_2^2 b_2$	+	$\Sigma x_2 x_3 b_3$	=	$\Sigma x_2 y$
R3	$\Sigma x_1 x_3 b_1$	+	$\Sigma x_2 x_3 b_2$	+	$\Sigma x_3^2 b_3$	=	$\Sigma x_3 y$

That is

$$X^T X \mathbf{b} = X \mathbf{y} \qquad (2.9.3)$$

Notice that

$$SSR = \mathbf{y}^T P \mathbf{y} = \mathbf{y}^T X (X^T X)^{-1} X^T \mathbf{y} = \mathbf{b}^T X^T \mathbf{y}$$

$$SSE = \mathbf{y}^T (I-P) \mathbf{y} = SST - SSR$$

The normal equations, along with SST, are all we need to test any hypothesis about the β's. For example

(i) H: $\beta_2 = \beta_3$ (This involves 1 d.f; $r = 1$)
The reduced model is

$$\mathbf{y} = \beta_1 \mathbf{x}_1 + \beta_2 (\mathbf{x}_2 + \mathbf{x}_3) + \boldsymbol{\varepsilon}$$

that is y is regressed on x_1 and (x_2+x_3). The reduced normal equations are:

$$\sum x_1^2 b_1 + \sum (x_2 + x_3) x_1 b_2 = \sum x_1 y$$

$$\sum (x_2 + x_3) x_1 b_1 + \sum (x_2 + x_3)^2 b_2 = \sum (x_2 + x_3) y$$

The appropriate quantities here could be obtained from (2.9.3) by adding C2 and C3 and also R2 and R3. The estimated coefficients will differ from those of the full model.

(ii) H: $\beta_2 = \beta_3 = 0$ (2 d.f; $r = 2$)
The reduced model is

$$\mathbf{y} = \beta_1 \mathbf{x}_1 + \boldsymbol{\varepsilon}$$

and the reduced normal equations can be formed from (2.9.3) by omitting columns C2 and C3 and rows R2 and R3.

(iii) H: $\beta_1 = 1$ ($r = 1$)
The reduced model is

$$\mathbf{y} = (1)\mathbf{x}_1 + \beta_2\mathbf{x}_2 + \beta_3\mathbf{x}_3 + \boldsymbol{\varepsilon}$$

$$\text{or} \quad \mathbf{y}-\mathbf{x}_1 = \beta_2\mathbf{x}_2 + \beta_3\mathbf{x}_3 + \boldsymbol{\varepsilon}$$

This is slightly different to the first two cases as $(y-x_1)$ is regressed on x_2 and x_3 causing the SST to change as well as SSR. The reduced normal equations are:

$$\sum x_2^2 b_2 + \sum x_2x_3b_3 = \sum x_2(y - x_1)$$
$$\sum x_2x_3b_2 + \sum x_3^2 b_3 = \sum x_3(y - x_1)$$

These quantities can be obtained from R2 and R3 with $b_1 = 1$.

We illustrate these three hypothesis with the following data.

Example 2.9.1 Heart data

The dependent variable, y, is the weight of a horse's heart measured at the time of a post-mortem. The three predictor variables are measurements taken by an ultra-sound device of the left ventricle when the horse was alive. The three measurements are the width of the ventricle inner wall, outer wall and exterior width during the diastole (relaxed) phase. The complete data set can be found in Appendix C.

The normal equations are

	C1		C2		C3		C4
R1	$19.850\ b_1$	+	$10.706\ b_2$	+	$72.222\ b_3$	=	27.242
R2	$10.706\ b_1$	+	$11.817\ b_2$	+	$51.393\ b_3$	=	17.780
R3	$72.222\ b_1$	+	$51.393\ b_2$	+	$364.566\ b_3$	=	109.264

The number of observations = 46 and SST = 56.846. Solving these equations yields

$b_1 = 0.971 \quad b_2 = 0.408 \quad b_3 = 0.050$ and

$SSR = 39.153$ so that $SSE = 17.693$

Note that the sum of squares for regression is found by multiplying each estimate by the corresponding element of the right hand side of the normal equations. Thus,

$$SSR = 0.971 \times 27.242 + 0.408 \times 17.780 + 0.050 \times 109.264$$

(i) H: $\beta_2 = \beta_3$

The reduced normal equations are

$$19.850\, b_1 + 82.928\, b_2 = 27.242$$
$$82.928\, b_1 + 479.169\, b_2 = 127.044$$

giving

$b_1 = 0.956 \quad b_2 = 0.100$

SSR(Reduced) = 38.706, SSE(Reduced) = 18.140

$$F_{1,42} = [0.447/1]/[17.693/42]$$
$$= 1.06$$ which is too small to reject H

(ii) H: $\beta_2 = \beta_3 = 0$

The reduced normal equation is

$$19.850\, b_1 = 27.242$$

or $b_1 = 1.372$

SSR(Reduced) = 37.387, SSE(Reduced) = 19.459

$$F_{2,42} = [1.766/2]/[17.693/42] = 2.096$$

Again this is too small to reject H.

(iii) H: $\beta_1 = 1$

$$11.817\, b_2 + 51.393\, b_3 = 17.780 - 10.706 = 7.074$$
$$51.393\, b_2 + 364.566\, b_3 = 109.264 - 72.222 = 37.042$$

$b_2 = 0.405 \quad b_3 = 0.044$

$$\text{SSR(Reduced)} = 0.405 \times 7.074 + 0.044 \times 37.042 = 4.495$$

$$\text{SST} = (y - x_1)^2 = y^2 + x_1^2 - 2x_1 y = 22.212$$

$$\text{SSE(Reduced)} = 17.717$$

$$F_{1,42} = [17.717 - 17.693]/[17.693/42] = 0.057$$

Hence H cannot be rejected at the 5% level.

2.10 PURE ERROR AND LACK OF FIT

When a model is fitted to data, its adequacy can be tested by the F-test of Section 2.5. If the model is correct, the residuals and the MSE, in particular, reflect the variation of the deviations. If the model is not the best one, the residuals and the MSE tend to be inflated; it can be shown that the MSE is still unbiased but its variance will increase if the model is not optimal.

If a number of observations of y are noted for the same set of values of the predictor variables, these repeated observations provide sums of squares and mean squares which are termed Pure Error. The mean square for Pure Error is an independent estimate of the variance of y and is unaffected by whichever model is fitted.

Consider the pH data of Appendix C2. We assume that for each value of x the three readings of y are independent which will be the case if the whole experiment has been repeated three times. (If the three readings are merely repeated measurements on the same sample we would expect them to be similar and as they are not independent they would not give a valid estimate of the variance of y.) When the time, x, is 4 the y values are 7.47, 7.49, 7.45 with a mean of 7.47. The contribution to the sum of squares for pure error is

$$0 + 0.02^2 + (-0.02)^2$$

and the degrees of freedom associated with this is 2. In similar fashion, the contribution from each of the other x values can be evaluated and summed to give SSPE. More formally, we can write

$$\text{SS Pure Error} = \sum_{ij}\sum (y_{ij} - \bar{y}_i)^2$$

The subscript j refers to the replications at the ith value of the predictor variables.

$$\text{SS(Lack of Fit)} = \text{SSE} - \text{SS(Pure Error)} \qquad (2.10.2)$$

The adequacy of the model is then tested by F = MS(LOF)/MS(PE) with the degrees of freedom appropriate to each of the mean squares. If F is larger than the critical value found from the tables then the model is rejected. Otherwise, the model is accepted and the MSE is adopted as an estimate of the variance of y for other tests and confidence intervals. Problems 2.5 and 2.6 explore this topic further.

2.11 EXAMPLE - LACTATION CURVE

In this problem, the quantity of milk (unit = 0.5 litres in a 24 hour period) was observed on one day a week for 38 weeks. The data refers to the first cow in Appendix C 3. The observations were (y,w) = (units of milk, week number). The lactation curve of y against w was to be modelled.

Some researchers have used a power curve to model this kind of data giving a model $y = \alpha\, e^{\beta\omega}$. Taking natural logarithms gives $\ln y = \ln \alpha + \beta w$. Other researchers have suggested the model referred to in Problem 2.3. We have added a quadratic term in w then tested it to see if this was significant. A computer package, MINITAB, was used to fit the model

$$\ln y = \beta_0 + \beta_1 w + \beta_2 w^2 + \varepsilon$$

Some of the output is given in Table 2.11.1. It is annotated to relate it to Chapters 1 and 2.

(i) The standard deviation of y about the regression line is $\sqrt{MSE} = 0.07961$

(ii) $R^2 = SSR/SST = 3.0676/3.2894$ or 93.3%

(iii) To test the hypothesis H: $\beta_2 = 0$ we can use the t-statistic = −5.24 with 35 degrees of freedom (the d.f. of s) or, we could use the partial F-statistic obtained from the sum of squares when t_2 is added to the model.

$$F_{1,35} = (SSR/1)/MSE = 0.1738/0.0063 = 27.59$$

Note that $(t_{35})^2 = (-5.24)^2 = 27.46$. F should equal t^2, which it does except for rounding error.

(iv) The variance of b_1 is $c_{11}\sigma^2$ where $(X^TX)^{-1}_{ij} = \{c_{ij}\}$. This is estimated by

$$\text{var } b_1 = c_{11}s^2 = 0.003686 \times 0.0063 = 0.00002322$$

The estimated standard error of b_1 is $\sqrt{0.00002322} = 0.0048$ as indicated by ST.DEV.OF COEF.

(v) The MINITAB program has printed out three points (rows 14, 31, 34) as these gave large studentized residuals. The estimated variance of the predicted value of y_{14}, the fourteenth observation, is given by

$$\text{var}(\hat{y}_{14}) = p_{14,14}\, s^2$$

TABLE 2.11.1 Lactation Curve Analysis

THE REGRESSION EQUATION IS
ln Y = 2.75 - 0.00062 w - 0.000629 w-squared

COLUMN	COEFFICIENT	ST. DEV. OF COEF.	T-RATIO = COEF/S.D.
	2.74952	0.04088	67.27
w	-0.000617	0.004833 (iv)	-0.13
w-squared	-0.0006294	0.0001202	-5.24 (iii)

S = 0.07961

R-SQUARED = 93.3 PERCENT (ii)
R-SQUARED = 92.9 PERCENT, ADJUSTED FOR D.F.

ANALYSIS OF VARIANCE

DUE TO	DF	SS	MS=SS/DF	
REGRESSION	2	3.0676	1.5338	
RESIDUAL	35	0.2218	0.0063	=s-squared (i)
TOTAL	37	3.2894		

FURTHER ANALYSIS OF VARIANCE
SS EYPLAINED BY EACH VARIABLE WHEN ENTERED IN THE ORDER GIVEN

DUE TO	DF	SS
REGRESSION	2	3.0676
w	1	2.8938
w-squared	1	0.1738 (iii)

ROW	C1	Y C10	PRED. Y VALUE	ST.DEV. PRED. Y	RESIDUAL	ST.RES.
14	14.0	2.4501	2.6175	0.0180 (v)	-0.1674	-2.16R (vi)
31	31.0	1.9601	2.1255	0.0188	-0.1654	-2.14R
34	34.0	2.2039	2.0009	0.0240	0.2030	2.67R

R DENOTES AN OBS. WITH A LARGE ST. RES.

DURBIN-WATSON STATISTIC = 1.75

where $\{p_{ij}\}$ is the projection matrix. Thus

$$p_{14,14}\, s^2 = (0.018)^2 \quad \text{giving}$$
$$p_{14,14} = (0.018)^2/0.0063 = 0.0514$$

The values of p_{ii}, and hence the var $\hat{y}$ will be smaller for the middle values of x. This point was made in Section 2.7 about the confidence intervals for E(y). The situation is somewhat more complicated with more than one predictor variable but the general rule holds that the variance of y will be larger for more extreme values of the predictor variables.

(vi) The estimated variance of the fourteenth residual is given by

$$\text{var } e_{14} = (1 - p_{14,14})s^2$$
$$= (1 - 0.0514)\times 0.0063 = 0.00598$$

The standard error of e_{14},

$$\text{s.e. } e_{14} = \sqrt{0.00598} = 0.0773$$

Hence, the fourteenth studentized residual is

$$\text{ST.RES.} = -0.1674/0.0773 = -2.16$$

As the 2.5th percentile of the t-statistic with 35 degrees of freedom is 2.03, this studentized residual falls outside of this and is tagged with the letter R, suggesting it may be an outlier.

(vii) The plot of studentized residuals against time in Figure 2.11.1 shows a few interesting features. There does seem to be some evidence for a mild correlation between residuals. The autocorrelation function confirms this. Figure 2.11.2. shows that by the first and second autocorrelations are positive but not large. The Durbin-Watson statistic is also slightly less than two.

The studentized residual of 2.67 also stands out suggesting that it may be an outlier, although it is not much larger than 97.5th

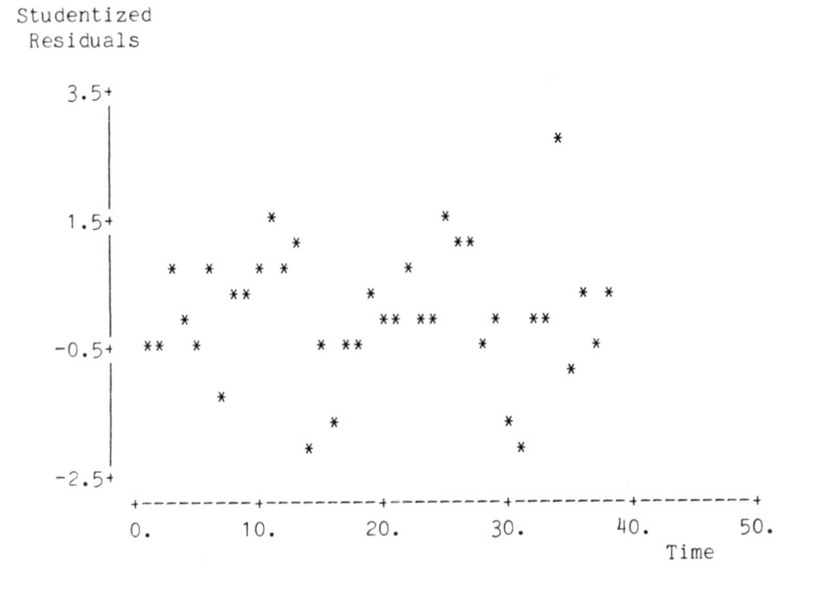

FIGURE 2.11.1 Studentized residuals plotted against time.

```
          -1.0 -0.8 -0.6 -0.4 -0.2  0.0  0.2  0.4  0.6  0.8  1.0
           +----+----+----+----+----+----+----+----+----+----+
 1   0.112                           XXXX
 2   0.216                           XXXXXX
 3  -0.306                   XXXXXXXXX
 4  -0.286                    XXXXXXXX
 5  -0.216                      XXXXXX
 6  -0.207                      XXXXXX
 7   0.147                           XXXXX
 8  -0.092                         XXX
 9   0.038                           XX
10  -0.120                        XXXX
11  -0.140                       XXXXX
12   0.033                           XX
13   0.059                           XX
14   0.235                           XXXXXXX
15   0.214                           XXXXXX
16   0.175                           XXXXX
```

FIGURE 2.11.2 Autocorrelation function of residuals.

percentile of the standard normal). With 38 residuals we could expect about 5% of 38 = 2 (approx) standard residuals to be greater in magnitude than this if the residuals followed a normal distribution. As there are only three there is nothing suspicious here.

The general shape of the residual plot suggests that the variance is approximately constant over time.

PROBLEMS

2.1 For the sports car data of Appendix C 4:-

(i) Plot the price against the year of manufacture. It seems that a quadratic model would be better than a linear one. Can you suggest a reason why this is so?

(ii) Show that the correlation between the year of manufacture and the square of this variable is close to 1. Why is this so and what problems could arise by using both of these variables as predictors with price as the dependent variable?

(iii) Show that writing the year in the form of deviations from its mean avoids the above problems.

(iv) What effect does the dealer have on the priced asked for a car? If the kind of dealer is included in the model, does this remove the need for the quadratic term used above?

2.2 For the data of Problem 1.5 (the results in 3 Math papers),

(i) Calculate the ANOVA table
(ii) Test the following hypotheses

(a) $\beta_1 = \beta_2 = 0$
(b) $\beta_1 = 0$

2.3 For the lactation curves in Section 2.11, some researchers prefer to fit ln y against w and ln w.

(i) Write a formula for this model in terms of y as a function of w.
(ii) Fit this model to the data in Appendix C 3 for cow no. 1.
(iii) Compare the fit of this model with that used in Section 2.11.

2.4 For quarterly data (observations made every three months) in which there are cyclical variations (for example, the December reading may tend to be high) explain what the ACF, autocorrelation function, would be like.

2.5 At each of m points, three observations of y are noted.

(i) Show that the model can be written as

$$\begin{vmatrix} \mathbf{y}_1 \\ \mathbf{y}_2 \\ \mathbf{y}_3 \end{vmatrix} = \begin{vmatrix} X \\ X \\ X \end{vmatrix} \boldsymbol{\beta} + \boldsymbol{\varepsilon}$$

Here, $\mathbf{y}_1$ is the m×1 vector of the first observation at each point, $\mathbf{y}_2$ is the vector of second observations, and $\mathbf{y}_3$ is that of the third. If there are k predictor variables and an intercept term, X is an m×(k+1) matrix.

(ii) Show that the estimator of $\boldsymbol{\beta}$ would be the same as in the model $\mathbf{y}^* = X\boldsymbol{\beta} + \boldsymbol{\varepsilon}$ where $\mathbf{y}^*$ is an m×1 vector of the average values of y at each of the m points. The residual sum of squares from this model is sometimes called the sum of squares for lack of fit.

(iii) As there are three values of y at each point, these can give a sum of squares for pure error. This enables us to firstly check whether the model is adequate by comparing the lack of fit with the pure error.

(a) Complete the ANOVA below for $m = 10$ and $k=5$.

Source	Degrees of Freedom	Sums of Squares	
Regression		24	
Residual		66	
Lack of fit			24
Pure error			42
Total		90	

(b) Test for no lack of fit.

2.6 For the pH data of Appendix C2:-

(i) Fit the model ln y against x, and also ln y against ln x. Which model do you recommend?

(ii) For the model above which you have chosen, test its adequacy by firstly evaluating the mean squares for lack of fit and pure error.

2.7 (i) Show that $K = I - (\mathbf{1}\ \mathbf{1}^T)/n$ is idempotent.

(ii) For the model $\mathbf{y} = \beta_0\mathbf{1} + \beta_1\ \mathbf{x}_1 + \boldsymbol{\varepsilon}_1$, where $\boldsymbol{\varepsilon}_1 \sim N(0, \sigma^2 I)$, show that multiplying through by K gives $\mathbf{y} = \beta\mathbf{x} + \boldsymbol{\varepsilon}_2$. where y and x are now in deviations from their means.

(iii) Show that $\varepsilon_2 \sim N(0, \sigma^2 K)$

(iv) If $P = P(\mathbf{x})$ is the projection matrix onto $\mathbf{x}$ show that MSR/MSE in Section 2.5 follows an F-distribution under H: $\beta = 0$. You will need to show that the expression can be written as a ratio of independent χ^2 distributions.

3 WHICH VARIABLES SHOULD BE INCLUDED IN THE MODEL

3.1 INTRODUCTION

When a model can be formed by including some, or all, of the predictor variables, there is a problem in deciding how many variables to include. The decision we arrive at will depend to some extent on the purpose we have in mind. If we merely wish to explain the variation of the dependent variable in the sample, then it would seem obvious that as many predictor variables as possible should be included . This can be seen with the lactation curve of Example 2.11. If enough powers of w were added to the model the curve would pass through every observed value, but it would be so jagged and complicated it would be difficult to understand what was happening. On the other hand, a small model has the advantage that it is easy to understand the relationships between the variables. Furthermore, a small model will usually yield estimators which are less influenced by peculiarites of the sample and so are more stable.

Another important decision which must be made is whether to use the original predictor variables or to transform them in some way, often by taking a linear combination. For example, the cost of a particular kind of fencing for a rectangular field may largely depend on the length and breadth of the field. If all the fields in the sample are in the same proportions then only one variable (length or breadth) would be needed. Even if they are not in the same proportions, one variable may be sufficient, namely the sum of the length and the breadth or, indeed, the perimeter. This is our ideal solution, reducing the number of predictor variables from two to one and at the same time obtaining a predictor variable which has physical meaning. With a particular data set, the predicted value of the cost may be $y = 1.1\,l + 0.9\,b$ so that the best single variable would be the right hand side with l = length and b = breadth, but this particular linear combination would have no physical meaning. We need to keep both aspects in mind, balancing statistical optimum against physical meaning.

In the first section we shall limit our discussion to orthogonal predictor variables. Although this may seem an unnecessarily strong restriction to place on the model, orthogonal variables often exist in experimental design situations. Indeed the values of the variables in the sample are often deliberately chosen to be orthogonal. We explain the advantages of this in Section 3.2, while in Section 3.4 we show that it is possible to transform variables, for any data set, so that they are orthogonal.

3.2 ORTHOGONAL PREDICTOR VARIABLES

If the variables in a model are expressed as deviations from their means and if there are k predictor variables, the sum of squares for regression is given by

$$\begin{aligned} SSR &= b_1 S_{y1} + b_2 S_{y2} + \cdots + b_k S_{yk} \\ &= S_{y1}^2 / S_{11} + S_{y2}^2 / S_{22} + \cdots + S_{yk}^2 / S_{kk} \qquad (3.2.1) \end{aligned}$$

The total sum of squares is

$$SST = S_{yy} = \sum y_i^2$$

By subtraction, we find the sum of squares for error (residual) is

$$SSE = SST - SSR \tag{3.2.2}$$

In this section, we assume that the predictor variables are orthogonal and explore the implications of the number of variables included in the model.

We consider now the effect of adding another variable, x_{k+1}, to the model and assume that this variable is also orthogonal to the other predictor variables. The SST will not be affected by adding x_{k+1} to the model. We introduce the notation that SSR(k) is the sum of squares for regression when the variables $x_1, x_2, \cdots, x_k$ are in the model. It is clear that

(i) $\quad SSR(k+1) \geqq SSR(k)$

This follows from (3.2.1) as each term in the sum cannot be negative so that adding a further variable cannot decrease the sum of squares for regression.

(ii) $\quad SSE(k+1) \leqq SSE(k)$

This is the other side of the coin and follows from (3.2.2).

(iii) $\quad R(k+1)^2 = SSR(k+1)/SST \geqq R(k)^2 = SSR(k)/SST$

SSR(k+1) can be thought of as the amount of variation in y explained by the (k+1) predictor variables, and $R(k+1)^2$ is the proportion of the variation in y explained by these variables. These monotone properties are illustrated by the diagrams in Figure 3.2.1.

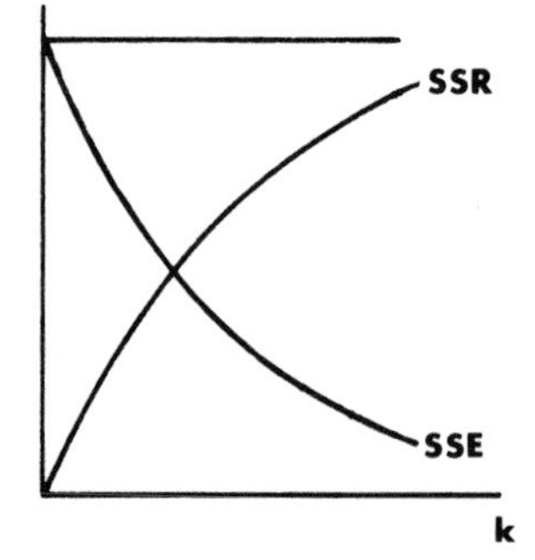

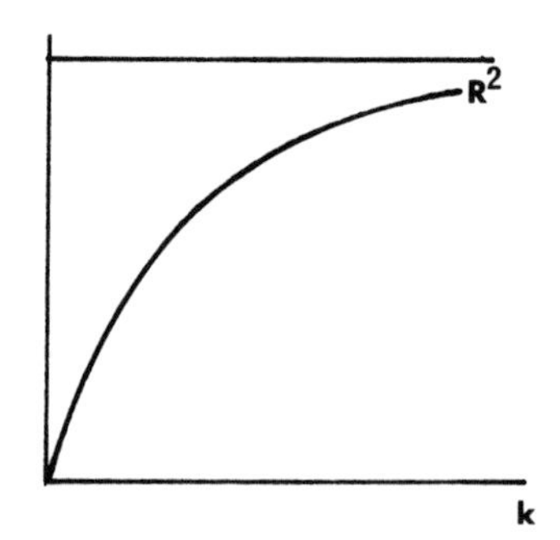

FIGURE 3.2.1 The general shapes of curves for SSR, SSE and R^2.

Two other statistics which are of interest in deciding how well a model fits are s^2, an estimate of σ^2, and the F-statistic for testing

$$H: \beta_1 = \beta_2 = \cdots = \beta_k = 0$$

Neither of these statistics exhibit the monotone increase (or decrease) of R^2, SSR and SSE.

(iv) $$s(k)^2 = SSE(k)/(n-1-k) = MSE(k)$$

As the number of variables, k, increases both the numerator and denominator decrease and s^2 will reach a minimum as illustrated in Figure 3.2.2.

(v) The F-statistic is given by

$$F_{k,\ n-k-1} = MSR(k)/MSE(k) \quad \text{where } MSR(k) = SSR(k)/k$$

Neither the numerator nor the denominator is monotone as k increases but F will attain a maximum and is illustrated in Figure 3.2.2.

The implications of the above discussion are that as orthogonal variables are added to the model more of the variation in the dependent variable is explained. For the testing of hypotheses, which involves s^2 and F, the addition of further variables may not improve the situation. Another point, although we have not demonstrated it, is that if the model is to be used for prediction, limiting the number of variables leads to more accurate prediction by reducing the variance of future predicted values. In Section 3.4, we show that

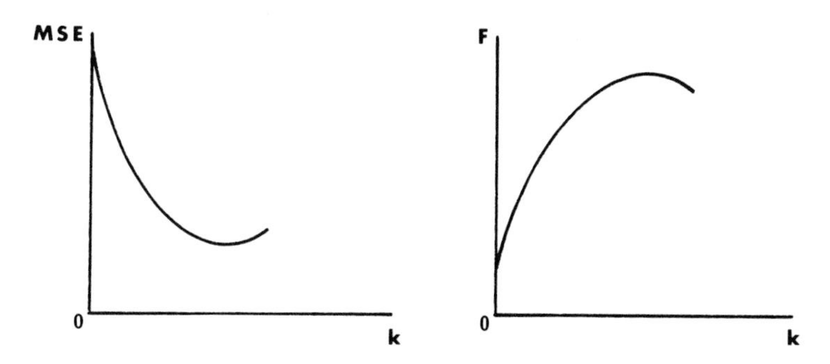

FIGURE 3.2.2 The variation of the mean square for residuals and the F-statistics with the number of predictors, k.

the above points apply also to nonorthogonal predictor variables. Orthogonal predictor variables represent the ideal situation statistically for the matrix

$$X^T X = \mathrm{diag}\{ S_{11} , S_{22} , \cdots, S_{kk}\}$$

with offdiagonal elements being zero. This means that the coefficient estimates are stable, independent of each other and the i-th estimate can be tested by

$$H: \beta_i = 0 \quad \text{using} \quad F = b_i^2 \; S_{ii} / s^2$$

Also, the sum of squares for regression is the sum of regression sum of squares of the individual regressions. This follows from the fact that the projection matrix

$$P = P_1 + P_2 + \cdots + P_k$$

3.3 LINEAR TRANSFORMATIONS OF THE PREDICTOR VARIABLES

If x_1 and x_2 are a student's marks on two tests during this year in a certain paper and y is the student's mark on the final exam (and for convenience we write these variables as deviations from their means), we may propose the model

$$\mathbf{y} = \beta_1\mathbf{x}_1 + \beta_2\mathbf{x}_2 + \boldsymbol{\varepsilon} \tag{3.3.1}$$

Alternatively, we could try to predict the final mark by transforming the scores of the two earlier tests to obtain two different variables, for example the sum of the two tests during the year and the difference between them.

$$\mathbf{w}_1 = \mathbf{x}_1 + \mathbf{x}_2 \quad \text{and} \quad \mathbf{w}_2 = \mathbf{x}_1 - \mathbf{x}_2.$$

The model is then

$$\mathbf{y} = \alpha_1\mathbf{w}_1 + \alpha_2\mathbf{w}_2 + \boldsymbol{\varepsilon} \tag{3.3.2}$$

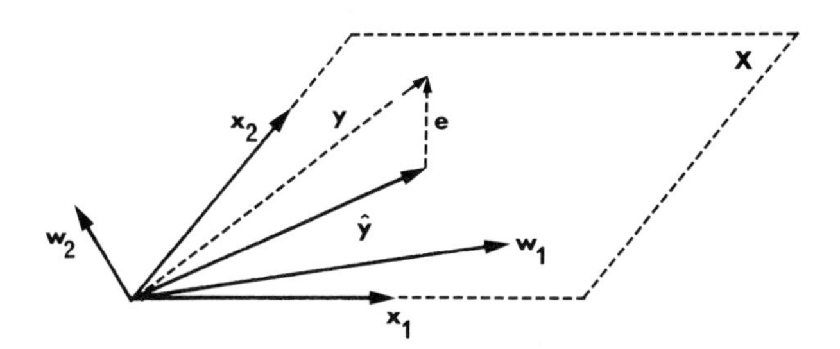

FIGURE 3.3.1 The projection of y on two predictor variables.

It is not difficult to show algebraically that $\hat{y}$, SSR and SSE are the same for each model. This can also be illustrated from a geometric viewpoint. The two **x** vectors define a plane X. From Appendix A, we see that $\mathbf{w}_1$ and $\mathbf{w}_2$ also lie in this plane and can be used as an alternative basis for it. This is illustrated by Figure 3.3.1 in which all the vectors, except y and e, lie in the plane X so that the predicted value, $\hat{y}$, is unaffected by the choice of a basis for X.

As we are free to choose any pair of vectors to define the plane X, another approach would be to choose two orthogonal vectors. We explore this is the next section.

3.4 ADDING NONORTHOGONAL VARIABLES SEQUENTIALLY

Although orthogonal predictor variables are the ideal, they will rarely occur in practice with observational data. If some of the predictor variables are highly correlated, the matrix X^TX will be nearly singular. This could raise statistical and numerical problems, particularly if there is interest in estimating the coefficients of the model. We have more to say on this in the next section and in a later section on Ridge Estimators.

Moderate correlations between predictor variables will cause few problems. While it is not essential to convert predictor variables to others which are orthogonal, it is instructive to do so as it gives insight into the meaning of the coefficients and the tests of significance based on them.

In Problem 1.5, we considered predicting the outcome of a student in the mathematics paper 303 (which we denoted by y) by marks received in the papers 201 and 203 (denoted by x_1 and x_2, respectively). The actual numbers of these papers are not relevant, but, for interest sake, the paper 201 was a calculus paper and 203 an algebra paper, both at second year university level and 303 was a third year paper in algebra.The sum of squares for regression when y is regressed singly and together on the x variables (and the R^2 values) are:

SSR on 201 alone :	1433.6	(.405)
SSR on 203 alone :	2129.2	(.602)
SSR on 201 and 203 :	2265.6	(.641)

Clearly, the two x variables are not orthogonal (and, in fact, the correlation coefficient between them is 0.622) as the individual sums of squares for regression do not add to that given by the model with both variables included. Once we have regressed the 303 marks on the 201 marks, the additional sum of squares due to 203 is (2265.6 - 1433.6) = 832. In this section we show how to adjust one variable for another so that they are orthogonal, and, as a consequence, their sums of squares for regression add to that given by the model with both variables included.

SSR for 201	= 1433.6	= SSR for x_1
SSR for 203 adjusted for 201	= 832.0	= SSR for z_2
SSR for 201 and 203	= 2265.6	

We start with the simple case of two predictor variables, x_1 and x_2, which are expressed as deviations from their means and we will assume they are not orthogonal to each other. We can make x_2 orthogonal to x_1 by adjusting x_2 for x_1 in a very natural way. We regress x_2 on x_1, and we call the residual z_2 as in Figure 3.4.1. As z_1 and z_2 define the same plane as x_1 and x_2, the predicted value of y, the dependent variable, remains the same regardless of which pair of predictor variables are used. We can write $\hat{y}$ as

$$\hat{\mathbf{y}} = b_1\mathbf{x}_1 + b_2\mathbf{x}_2 = a_1\mathbf{z}_1 + a_2\mathbf{z}_2$$

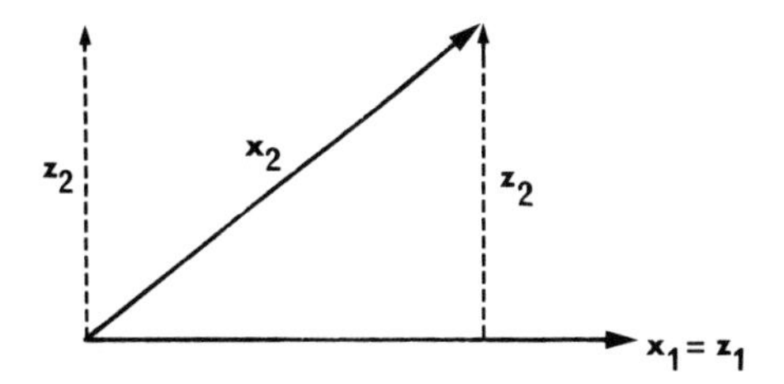

FIGURE 3.4.1 Adjusting one predictor variable, x_2, for another, x_1.

Two interesting points should be noted

(i) $b_2 = a_2$

This can easily be shown by writing

$$\mathbf{z}_2 = (I-P)\,\mathbf{x}_2 = \mathbf{x}_2 - \mathbf{x}_1(\mathbf{x}_1^T\mathbf{x}_1)^{-1}\mathbf{x}_1^T\mathbf{x}_2$$
$$= \mathbf{x}_2 - A\,\mathbf{x}_1$$

Thus,

$$a_1\mathbf{z}_1 + a_2\mathbf{z}_2 = a_1\mathbf{z}_1 + a_2\mathbf{x}_2 - a_2\,A\,\mathbf{x}_1$$
$$= (a_1 - a_2A)\,\mathbf{x}_1 + a_2\mathbf{x}_2$$

As this must equal $b_1\mathbf{x}_1 + b_2\mathbf{x}_2$, we have

$$a_2 = b_2 \quad \text{and} \quad b_1 = a_1 - a_2\,A$$

In words, the estimate b_2 is the same as if x_2 was added last to the model and adjusted for those variables already in the model.

(ii) The predicted value of y and the sum of squares for regression is the same for the orthogonal z variables as the correlated x variables. This occurs because the predicted value of y is the projection onto the plane defined by the x's which is the same as the plane defined by the z's.

$$SSR = \hat{\mathbf{y}}^T\,\hat{\mathbf{y}}$$

Furthermore, SSR can be expressed in terms of the projection matrix P.

$$SSR = (P\,\mathbf{y})^T(P\,\mathbf{y}) \quad \text{where } P = P_1 + P_2$$

where P_i is the projection onto z_i. Clearly we could add another variable x_3 and transform it to z_3 which is x_3 adjusted for x_1 and x_2. As

$$SSR(\text{for } x_1, x_2, x_3) = SSR(z_1, z_2, z_3)$$
$$= SSR(z_1) + SSR\,(z_2) + SSR\,(z_3)$$

and the z_i are orthogonal to each other, the five properties, (i) through (iv), of Section 3.2 also hold for the z_i (and consequently the nonorthogonal x_i). The sums of squares for the adjusted variables, z_i, are called the sequential sums of squares.

The biggest drawback to transforming to orthogonal vectors is that the values depend on the order that the variables are added to the model. It does, however, shed light on the least squares process and the meaning of the estimated coefficients.

3.5 CORRELATION FORM

When the main concern is to decide which variables to include in the model, a very useful transformation of the data is to scale each variable, predictors and dependent variables alike, so that the normal equations can be written in correlation form. This enables us to identify important variables which should be included in the model and it also reveals some of the dependencies between the predictor variables.

As usual, we consider the variables to be in deviation form. The correlation coefficient between x_1 and x_2 is

$$r_{12} = S_{12} / \sqrt{(S_{11}\ S_{22})} = \sum x_1\ x_2 / \sqrt{(S_{11}\ S_{22})} \tag{3.5.1}$$

If we divide each variable x_i by $\sqrt{S_{ii}}$ and denote the result as

$$x_i^* = x_i / \sqrt{S_{ii}}$$

then x^*_i is said to be in correlation form. Notice that

$$\begin{aligned} &\text{(i)} && \Sigma\, x_i^* = 0 \\ &\text{(ii)} && \Sigma (x_i^*)^2 = 1 \\ &\text{(iii)} && \Sigma\, x_i^*\, x_j^* = r_{ij} \end{aligned} \tag{3.5.2}$$

We have transformed the model from

$$y = \beta_1 x_1 + \beta_2 x_2 + \varepsilon \quad \text{to} \quad y^* = \alpha_1 x_1^* + \alpha_2 x_2^* + \varepsilon$$

and the normal equations simplify from

$$\begin{aligned} S_{11} b_1 + S_{12} b_2 &= S_{y1} \\ S_{12} b_1 + S_{22} b_2 &= S_{y2} \end{aligned} \quad \text{to} \quad \begin{aligned} a_1 + r_{12} a_2 &= r_{y1} \\ r_{12} a_1 + a_2 &= r_{y2} \end{aligned} \qquad (3.5.3)$$

It is generally true that if the correlation of x_i with y is large then the coefficient a_i tends to be large and the variable x_i is important in the model. This is obvious in the special case where x_1 and x_2 are orthogonal. Then

$$a_i = r_{yi}$$

$$SSR = a_1 r_{y1} + a_2 r_{y2} = r_{y1}^2 + r_{y2}^2$$

On the other hand, a high correlation between two predictor variables would suggest that both variables would not be needed in the model as they each explain more or less the same variation in y. Furthermore, if the coefficients are themselves of interest, then a high correlation between them inflate the variance of the estimates. This can be seen with a model with two predictor variables.

$$\text{var} \begin{vmatrix} a_1 \\ a_2 \end{vmatrix} = \begin{vmatrix} 1 & r_{12} \\ r_{12} & 1 \end{vmatrix}^{-1} \sigma^2 = \frac{1}{1 - r_{12}^2} \begin{vmatrix} 1 & -r_{12} \\ -r_{12} & 1 \end{vmatrix} \sigma^2$$

That is

$$\text{var } a_1 = \text{var } a_2 = \sigma^2 / (1 - r_{12}^2)$$

As the correlation increases towards 1, then the variances of the estimates increase without limit. The estimated coefficients then become unstable. Notice that the determinant of the $X^T X$ matrix is

$$\det \begin{vmatrix} 1 & r_{12} \\ r_{12} & 1 \end{vmatrix} = 1 - r_{12}^2$$

so as the correlation tends to 1, the determinant tends to 0. If the determinant equals zero the matrix is singular and its inverse does not exist. Even if the determinant is just near to zero, computational problems arise which could produce garbage. For more than two predictor variables, the same ideas hold and large values of correlations between predictor variables tend to make the determinant small which could lead to problems in solving the normal equations.In recent years much consideration has been given to the numerical solution of linear equations such as the normal equations so that with most computer programs using sophisticated algorithms the major problem may be in the statistical interpretation of the results rather than the numerical problems of obtaining a solution.

It is worthwhile studying the correlation matrix in some detail as it is a very good starting point for understanding the relationships between the variables. The following example illustrates some of these points.

Example 3.5.1 Heart data

In the horse's heart's data of Appendix C1 the set of three measurements was also made during the systolic (contracted) phase. The predictor variables obtained by the ultrasound device were therefore

x_1, x_3, x_5 : widths of the ventricle inner wall, outer wall and exterior width during the systole phase and

x_2, x_4, x_6 : the same measurements during the diastole phase

The number of observations, n = 46 and the correlation matrix is

		x_2	x_3	x_4	x_5	x_6	y
A	x_1	0.909	0.825	0.756	0.877	0.807	0.778
B	x_2		0.772	0.699	0.812	0.849	0.811
C	x_3			0.908	0.749	0.792	0.779
D	x_4				0.724	0.783	0.686
E	x_5					0.961	0.681
F	x_6						0.759

For convenience, the variables x_1 - x_6 are also labelled A-F. The matrix is symmetric as $r_{ij} = r_{ji}$ so that only half the matrix is printed. As $r_{ii} = 1$ the diagonals are omitted.

A number of facts about the data emerge from the correlation matrix. All of the correlation coefficients are positive and reasonably large which indicates that with large hearts all the lengths increase in a fairly uniform manner. The predictor variables are highly correlated, particularly between the two phases of the same length (A and B, C and D, E and F). This suggests that not all of these variables are needed but only a subset of them. Indeed, the high correlations should make us hesitant to make unique claims about any particular subset.

The largest correlation is y with B, 0.811, so that individually B has the greatest influence on y, followed by C and A . Of course, some relationships will not be clear from the correlations. For example, a strong correlation between B and C may be due to the fact that both have a high correlation with a third variable, say Z. In this case, if B is adjusted for Z and C is also adjusted for Z then the correlation between the adjusted B and C may be low. For example suppose that observations are taken annually and

Z is the population of New Zealand
B is the number of meat pies sold
C is the number of burglaries reported.

During a time of high growth in population, the number of pies and burglaries would probably also increase. The correlation between B and C may be large and positive in this case, suggesting a strong but inexplicably close relationship between these two variables. The close agreement may merely be due to the fact that these two variables have increased in a time of high growth in population. That is, an increase in Z has led to an increase in both B and C . Some authors have described the resulting high value of the correlation as being due to the "lurking variable" Z.

These ideas form the rationalisation for the backward elimination and stepwise methods of arriving at an optimum subset of predictor variables in the model. Unfortunately, different methods can lead to different models. Fortunately, the solutions are often similar in that either of two highly correlated variables may be included in the model.

3.6 VARIABLE SELECTION - ALL POSSIBLE REGRESSIONS

In many situations, researchers know which variables may be included in the predictor model. There is some advantage in reducing the number of predictor variables to form a more parsimonious model. One way to achieve this is to run all possible regressions and to consider such statistics as the coefficient of determination, $R^2 = SSR/SST$.

We will use the heart data of Section 3.5, again relabelling the variables as A through F. With the variables in correlation form, $R^2 = SSR$, the sum of squares for regression, and this is given for each possible combination of predictor variables in Table 3.6.1.

TABLE 3.6.1 SSR For Each Possible Regression For the Heart Data

p = 2	3	4	5	6	7
A .605	AB .667	ABC .716	ABDC .718	ABCDE .719	ABCDEF .753
B .658	BC .715	BCD .718	BCDE .719	BCDEF .749	
C .606	CD .609	CDE .633	CDEF .703	CDEFA .742	
D .471	DE .542	DEF .621	DEFA .710	DEFAB .723	
E .463	EF .607	EFA .709	EFAB .722	EFABC .743	
F .576	FA .655	FAB .681	FABC .717	FABCD .721	
	AC .664	ACD .667	ACDE .669		
	BD .686	BDE .686	BDEF .712		
	CE .628	CEF .684	CEFA .725		
	DF .598	DFA .658	DFAB .690		
	EA .613	EAB .667	EABC .718		
	FB .676	FBC .717	FBCD .721		
	AD .628	ADE .630	ADEB .688		
	BE .659	BEF .704	BEFC .741		
	CF .660	CFA .684	CFAD .693		
		DAB .687			
		EBC .717			
		FCD .673			
		AEC .665			
		BFD .689			

To assist the choice of the best subset, C.L. Mallows suggested fitting all possible models and evaluating the statistic

$$C_p = SSE_p/s^2 - (n-2p) \tag{3.6.1}$$

Here, n is the number of observations and p is the number of predictor variables in the subset, including a constant term. For each subset, the value of Mallows' statistic can be evaluated from the correponding value of SSR. The complete set of these statistics are listed in Table 3.6.2. For each subset we use the mean squared error, MSE, of the full model as an estimate of the variance.

Suppose that the true model has q predictor variables. Thus

$$\mathbf{y} = X_q \boldsymbol{\beta}_q + \varepsilon \qquad \text{or} \qquad \mathbf{y} \sim N(X_q\beta_q, \sigma^2 I)$$

However, suppose that the fitted model includes p variables. The

TABLE 3.6.2 Mallows' Statistic Values For Subsets of Predictor Variables For the Heart Data

P = 2	3	4	5	6	7
A 20.3	AB 12.5	ABC 6.8	ABCD 8.4	ABCDE 10.3	ABCDEF 6.9
B 11.9	BC 4.9	BCD 6.4	BCDE 8.3	BCDEF 5.6	
C 20.1	CD 21.6	CDF 19.8	CDEF 10.8	CDEFA 6.7	
D 41.4	DE 32.0	DEF 21.7	DEFA 9.7	DEFAB 9.7	
E 42.6	EF 21.9	EFA 7.9	EFAB 7.8	EFABC 6.5	
F 24.8	FA 14.4	FAB 12.3	FABC 8.6	FABCD 10.0	
	AC 13.0	ACD 14.5	ACDE 16.2		
	BD 9.5	BDE 11.5	BDEF 9.4		
	CE 18.6	CEF 11.5	CEFA 7.3		
	DF 23.4	DFA 15.9	DFAB 12.9		
	EA 21.0	EAB 14.5	EABC 8.4		
	FB 11.1	FBC 6.6	FBCD 8.0		
	AD 18.6	ADE 20.3	ADEB 13.2		
	BE 13.7	BEF 8.7	BEFC 4.8		
	CF 13.6	CFA 11.8	CFAD 12.4		
		DAB 11.3			
		EBC 6.6			
		FCD 13.5			
		AFC 14.8			
		BFD 11.0			

projection matrix is $P_p = X_p(X_p^TX_p)^{-1}X^T$, the vector of residuals is $\mathbf{e}_p = (I-P_p)\mathbf{y}$ and the sum of squares for residuals is $SSE_p = \mathbf{e}_p{}^T\mathbf{e}_p$.

$$E(\mathbf{e}_p) = (I-P_p)E(\mathbf{y}) = (I-P_p)\, X_q\, \boldsymbol{\beta}_q = \text{bias} \qquad (3.6.2)$$

If we have fitted the true model, $p = q$, $(I-P_p)X_q = 0$ and $E(\mathbf{e}_p) = 0$ showing that the residuals are unbiased.

From Property 3 of Appendix B 3,

$$E(SSE_p) = \text{trace } (I-P_p)\,\sigma^2 + B_q^T\, X_q^T\, (I-P_p)\, X_q\, \boldsymbol{\beta}_q \qquad (3.6.3)$$

$$= (n-p)\sigma^2 + \text{bias}^T\text{ bias}$$

Now, Mallows' $C_p = SSE_p/s^2 - (n-2p)$. If the true model has been fitted

$$E(C_p) = (n-p)\sigma^2/\sigma^2 - (n-2p) = p$$

so that we expect C_p to be close to the diagonal line in Figure 3.6.1. This result is only approximate as we had to estimate the variance in the denominator. If the true model has not been fitted, p is not equal to q, and Mallows' statistic is inflated by the bias term of (3.6.3) which is always greater than, or equal to, zero. In

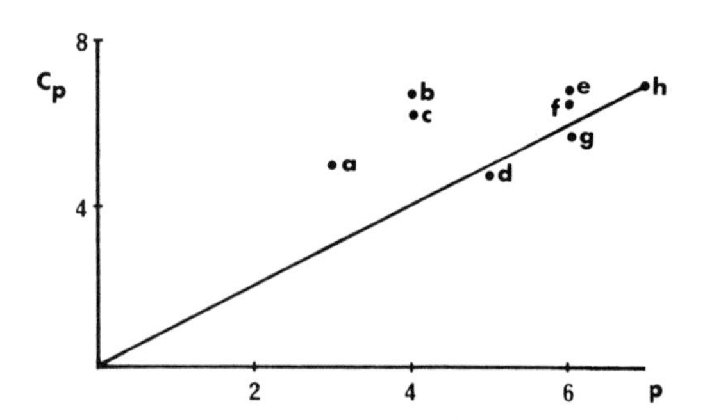

FIGURE 3.6.1 Mallows' C statistic for the Heart Data. The letters on the graph refer to the subsets of predictor variables as follows: a BC; b FBC or EBC; c BCD; d BEFC; e CDEFA; f EFABC; g BCDEF; h the full model ABCDEF.

practice, we assume the full model is unbiased as we use it to obtain the variance estimate.

From Figure 3.6.1, the subsets which are close to the diagonal line and which, consequently, are likely to have small bias are:

BC, BCEF, BCDEF, ACDEF, ABCEF, and the full model ABCDEF.
a d g e f h

The model BCEF would best satisfy the criteria of parsimony and small bias.

3.7 VARIABLE SELECTION - SEQUENTIAL METHODS

When the number of possible variables in a model is large, it may be inappropriate to run every possible regression and evaluate Mallows' statistic for each one, even though short cuts can be taken to evaluate such statistics by adding or subtracting terms rather than by evaluating each one from scratch.

Another approach is to add, or remove, variables, sequentially. We have seen that adding a variable will increase SSR, the sum of squares for regression. From Section 3.4 we could perform an F-test to decide if the increase in SSR is significant. The first method we consider is that of forward selection.

3.7.1 Forward Selection

We shall use the heart data of the last two sections to illustrate this. In Section 3.5, this data is written in correlation form.

If the model is to include only one predictor variable, then B would be chosen as it gives the highest SSR which is also the correlation coefficient with y. Before B is placed in the model, we test that it has a significant effect on y by using an F-test, or equivalently, a t-test.

We test H: $\beta_2 = 0$ in the model $y = \beta_0 + \beta_2 x_2 + \varepsilon$

$$F = (SSR/1) / (SSE/44) = 0.657 / (0.342/44) = 84.7$$

Clearly, we reject H and include x_2 in the model. We now try to add another predictor variable to the model. We look for the variable which, with B, gives the highest value of SSR. From Table 3.6.1, we see that SSR for B and C = 0.715 which is greater than for AB = 0.667, BD = 0.686, FB = 0.676, and BE = 0.659. Does C add significantly to SSR over and above B itself? We use the method of reduced models to determine this.

Full model: $y = \beta_0 + \beta_2 x_2 + \beta_3 x_3 + \varepsilon$; SSE = 0.285

Reduced model: $y = \beta_0 + \beta_2 x_2 + \varepsilon$; SSE = 0.342

Difference = 0.057

$$F = 0.057 / (0.285/43) = 8.6$$

The tabulated F at 5% level with 1,43 degrees of freedom = 4.1 so that we reject the reduced model in favor of the full model.

We then attempt to add a third variable to the model. The variable which adds most to SSR in association with B and C is D as BCD gives an SSR = 0.718. An F statistic is evaluated to determine whether D adds significantly to SSR over and above B and C.

Full model: $y = \beta_0 + \beta_2 x_2 + \beta_3 x_3 + \beta_4 x_4 + \varepsilon$; SSE = 0.282

Reduced model: $y = \beta_0 + \beta_2 x_2 + \beta_3 x_3 + \varepsilon$; SSE = 0.285

Difference = 0.003

$$F = 0.003 / (0.282/42) = 0.4$$

Clearly this is too small to reject the reduced model and we select as the optimal model that with B and C as predictor variables.

3.7.2 Backward Elimination

Another approach is to commence with the full model of six predictor variables and to attempt to remove variables sequentially. In Table 3.6.1, the five predictor model with the greatest SSR is BCDEF = 0.749 or SSE = 0.251. To decide if A should be removed, we compare this SSE with that of the full model using the F statistic.

$$F = (0.251-0.247) / (0.247/39) = 0.004 / (0.247/39) = 0.6$$

Clearly, the effect of A is not significant and can be removed. We look to remove one of these remaining five variables by considering the SSR for each of the four predictor models. These are:

BCDE = 0.719, CDEF = 0.703, BDEF = 0.712, FBCD = 0.721
and BEFC = 0.741

We choose this last one with D omitted and test whether this causes a significant reduction in SSR. The full model is now BCDEF and the reduced model is BCEF.

$$F = 0.008 / (0.251/40) = 1.3$$

This value is low compared with the 5% tabulated value for 1 and 40 degrees of freedom which equals 4.08 so that we proceed to eliminate a further variable. The three variable sums of squares are

CEF = 0.684, FBC = 0.717, BEF = 0.704, EBC = 0.717

For either of the models with SSR = 0.717,

$$F = 0.024 / (0.259/41) = 3.8$$

This is slightly below the critical value of 4.08 so that we proceed and compare these two models, FBC and EBC, with BC the last subset of these with two variables.

$$F = 0.002 / (0.283/42) = 0.3$$

We are then reduced to the model BC as in the forward selection process.

There are a number of points to notice about these sequential methods.

(i) In this particular case, the forward and backward procedures led to the same model. At times, this will not be the case and a choice will be necessary between models. We have more to say on such a choice in (iv) and (v) below.

(ii) The backward elimination procedure at one step could not choose between the two models EBC and FBC, which is not very surprising as the correlation coefficient for E and F is highly positive at 0.961. Either model would have been equally effective, but in this case no decision had to be made as the procedure moved on to the next step.

(iii) An arbitrary decision was made for both procedures to use the 5% level for deleting and adding variables. Other percentage levels could be used, or alternatively, a particular value of F. Practitioners in specific areas would, no doubt, develop a feel for the level appropriate for the kind of data they encounter.

(iv) The sequential methods described so far have limitations, in that, whereas the best single predictor variable may be C, the best pair of predictor variables may, in fact, be A and D. Draper and Smith (1981) suggest two possible approaches to this problem. The first is to run a stepwise regression which may select (say) 4 variables. Before accepting this model, the researcher would fit all the other models comprising combinations of 4 predictor variables and choose the best of these, where best could mean having the largest R-squared value. Alternatively, the researcher could increase the probability of introducing more variables into the model by reducing the entry level for forward selection, or increasing the rejection level for backward selection. This larger model is more likely to contain the most influencial variables, and, if it was thought appropriate, some of these could then be culled from the model.

(v) Other factors may enter into a decision on which variables to include in the model. Residual plots for some of the better models may indicate that, whereas one model fits very well in a certain area of the sample space, another model may give a better overall fit or a better fit under circumstances where it will be used to predict future values. Besides the purely statistical considerations, there may be other pressures to include particular variables. In an economic model, for example, the statistically optimal model may include an inflation index but not a specific time variable because the index may have been highly correlated with time. It may be politic, however, to also include the time variable to add credibility to the model, providing, of course, that its associated negative effects are not considerable.

3.8 QUALITATIVE (DUMMY) VARIABLES

It is often useful to introduce variables into a model to enable certain specific effects to be revealed and tested. Usually these take the form of qualitative variables which show up the differences between subgroups in the data. We shall use an example to explore these ideas.

In Example 1.5.1, we listed the value of an Australian stamp (1963 twopenny sepia coloured in the years 1972-1980). We could compare this with the listed value of another stamp, and for few obvious reasons, we have chosen the 1867 New Zealand fourpenny rose colored full face queen. We shall use the same transformation as before, namely

$$y_1 = \ln v_t,\ y_2 = \ln v_t$$

for the Australian and New Zealand stamp respectively. The data is given in Table 3.8.1. We could fit a separate model to each stamp, that is, for the Australian stamp

$$\mathbf{y}_1 = \alpha_1 \mathbf{1} + \alpha_2 \mathbf{t}_1 + \boldsymbol{\varepsilon}_1$$

and the New Zealand stamp

$$\mathbf{y}_2 = \beta_1 \mathbf{1} + \beta_2 \mathbf{t}_2 + \boldsymbol{\varepsilon}_2 \tag{3.8.1}$$

If the distributions of the deviations can be assumed to be the same, it will be advisable to join these models into a single model. The reason for this is that the sample size will be larger than for the

TABLE 3.8.1 Values of Australian and NZ Stamps

Year	t	Australian Value	NZ Value	Transformed values y_1	y_2
1972	0	10	22	0.	0.
1973	1	12	24	0.182	0.087
1974	2	12	28	0.182	0.241
1975	3	22	42	0.788	0.647
1976	4	25	50	0.916	0.821
1977	5	45	70	1.504	1.157
1978	6	75	150	2.015	1.920
1979	7	95	200	2.251	2.207
1980	8	120	600	2.481	3.306

individual models and large sample size will generally lead to more precise estimates of residual variance. This ensures that tests are more sensitive at discerning differences where they exist. This is achieved as follows

$$\begin{vmatrix} \mathbf{y}_1 \\ \mathbf{y}_2 \end{vmatrix} = \begin{vmatrix} \mathbf{1} & \underline{0} & \mathbf{t}_1 & \underline{0} \\ \underline{0} & \mathbf{1} & \underline{0} & \mathbf{t}_2 \end{vmatrix} \begin{vmatrix} \alpha_1 \\ \beta_1 \\ \alpha_2 \\ \beta_2 \end{vmatrix} + \begin{vmatrix} \varepsilon_1 \\ \varepsilon_2 \end{vmatrix}$$

$$\text{or} \quad \mathbf{y} = X\boldsymbol{\delta} + \varepsilon \qquad (3.8.2)$$

We shall refer to this as the full model. X is an 18×4 matrix. The first two vectors in X indicate whether the y observation relates to the Australian or New Zealand stamp. They may seem somewhat artifical, and hence the name "dummy". They point to qualitative differences rather than quantitative differences as in the case of y and t. In this particular example, we could drop the subscripts from the t vectors as they are identical for the two stamps.

We can perform a number of tests using the full model and appropriate reduced models. The ANOVA for each model is included at the end of this section.

(i) H: $\alpha_1 = \beta_1$

that is, the intercepts are the same for each stamp. The reduced model is

$$\begin{vmatrix} \mathbf{y}_1 \\ \mathbf{y}_2 \end{vmatrix} = \begin{vmatrix} \mathbf{1} & \mathbf{t}_1 & \underline{0} \\ \mathbf{1} & \underline{0} & \mathbf{t}_2 \end{vmatrix} \begin{vmatrix} \alpha_1 \\ \alpha_2 \\ \beta_2 \end{vmatrix} + \varepsilon \qquad (3.8.3)$$

$$\begin{aligned} F_{1,14} &= [\text{SSE(REDUCED)} - \text{SSE}]/[\text{SSE}/14] \\ &= [1.1104-1.0573]/[1.0573/14] \\ &= 0.70 \end{aligned}$$

(ii) H: $\alpha_1 = \beta_1$

that is, the slopes are the same. The reduced model is

$$\begin{vmatrix} \mathbf{y}_1 \\ \mathbf{y}_2 \end{vmatrix} = \begin{vmatrix} \mathbf{1} & \underline{0} & \mathbf{t}_1 \\ \underline{0} & \mathbf{1} & \mathbf{t}_2 \end{vmatrix} \begin{vmatrix} \alpha_1 \\ \beta_1 \\ \alpha_2 \end{vmatrix} + \varepsilon \qquad (3.8.4)$$

$$\begin{aligned} F_{1,14} &= [1.1286-1.0573]/[1.0573/14] \\ &= 0.94 \end{aligned}$$

(iii) H: $\alpha_1 = \beta_1$ and $\alpha_2 = \beta_2$
that is, the slopes and the intercepts are equal, or the same prediction line fits both data sets. The reduced model is

$$\begin{vmatrix} \mathbf{y}_1 \\ \mathbf{y}_2 \end{vmatrix} = \begin{vmatrix} \mathbf{1} & \mathbf{t}_1 \\ \mathbf{1} & \mathbf{t}_2 \end{vmatrix} \begin{vmatrix} \beta_1 \\ \beta_2 \end{vmatrix} + \varepsilon \qquad (3.8.5)$$

$$F_{2,14} = [(1.288-1.0573)/2]/[1.0573/14]$$
$$= 1.53$$

As the t values are the same for each stamp, the reduced model could be written as

$$y = \beta_1 + \beta_2 t + \varepsilon \qquad (3.8.6)$$

As $F_{1,14}(.95) = 4.60$ and $F_{2,14}(.95) = 3.74$ none of these hypotheses could be rejected and we would take our model as (3.8.5). In other words, there is every reason to aggregate these two sets of data into the one model.

Dummy variables can also be employed to decide whether the slope of a line has changed over time. When the values of the two stamps are plotted against time there appears to be a break in the line between t=5 and t=6 (1977 and 1978). To decide if two lines, as shown in Figure 3.8.1, would fit better than the single line of the reduced model, dummy variables could be introduced in a similar way to the full model of (3.8.1), in which the transposes of

$$\mathbf{t}_1 = (0,0,1,1,\cdots,5,5)^T$$

$$\mathbf{t}_2 = (6,6,7,7,8,8)^T$$

$$\mathbf{y}_1 = (0,0,\cdots,1.504,1.157)^T$$

$$\mathbf{y}_2 = (2.015,1.920,\cdots,2.481,3.306)^T$$

In this example the slope(s) of the line(s) would be of interest to a philatelist who had acquired the stamps for investment purposes. This may not be very easy to translate into practical terms, however, due to the logarithmic transformation. This again points to the advantages and disadvantages of transformations for, in this case, the transformation has led to a model with simple structure but its parameters are not amenable to easy understanding.

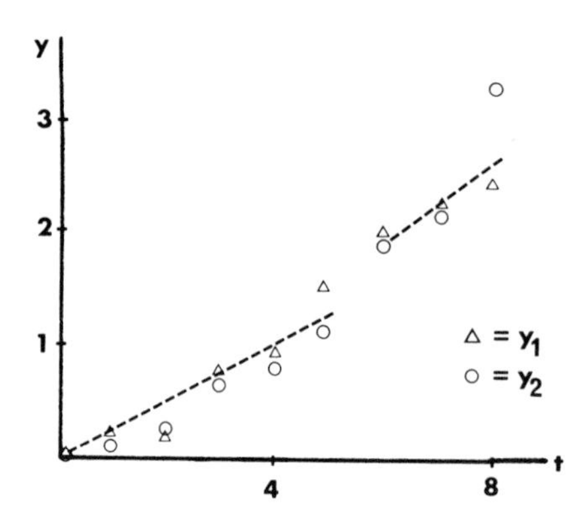

FIGURE 3.8.1 Values of y_1 and y_2, of two stamps over time.

3.9 AGGREGATION OF DATA

The dummy variable approach of the previous section provides one method of combining together different sets of data. In other situations it may be appropriate to merely sum or average data sets, but it is wise to tread warily for the pattern followed by an individual data set may be quite different from that of the aggregate of the data sets. Aggregation sometimes gives rather disturbing and nonsensical results. Consider Figure 3.9.1 in which there is a slight overall downwards trend in the y values as the x values increase. If there are, as shown, three identifiable subgroups in the data, they may, in fact, suggest a positive trend within each group. In this example, it would be spurious to aggregate the groups. One overall model could be used provided that dummy variables were included to distinguish the y-intercepts of each group.

For the lactation records of Appendix C 3, it would be useful to fit an overall model to the aggregated data of the five cows. One approach may be to fit a model to each cow's yield and then to average the estimated coefficients. Problems remain, however, for the model with these averaged coefficients may not fit well any of

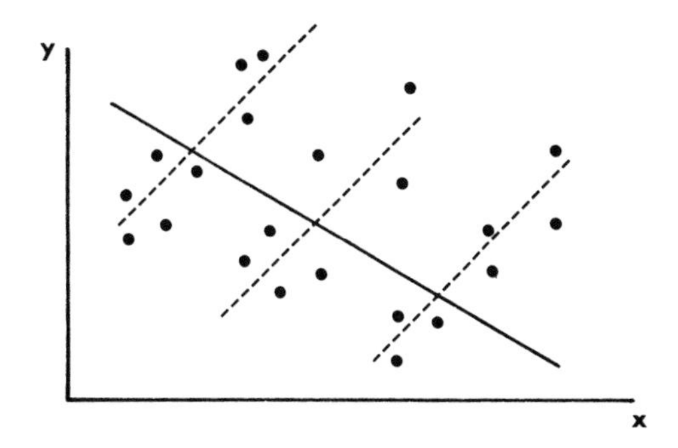

FIGURE 3.9.1 Aggregation of data can lead to spurious results if subgroups are present.

the individual data sets. Another approach is to average the sets of values of the dependent variable. With different patterns of yields, one must ask whether this aggregation is a reasonable thing to do. For example, the maximum yield for cow no. 1 occurred in the sixth week, but for cow no. 5 it occurred in the ninth week. Perhaps the data for each cow should be lagged so that the maxima correspond, and then the appropriate milk yields added (that is 16.30 of cow no. 1 added to 31.27 of cow no. 2 etc.). This would ensure that the maxima correspond, but it does not take into account the different shapes of the graphs or that one cow may give milk for a longer time than another. (For convenience, the lactation records here were all truncated to 38 weeks even though some actually gave milk for a longer period).

When ln y (logarithms of the milk yields) are averaged over the five data sets and regressed against w and ln w (where w is the week number) the residuals follow a highly cyclical pattern which is clearly shown in Figure 3.9.2 and is confirmed by the ACF, autocorrelation function, of the residuals in Figure 3.9.3. This cyclical pattern is quite surprising and suggests weaknesses in the model - perhaps additional predictor variables should be added or a weighted least squares model used.

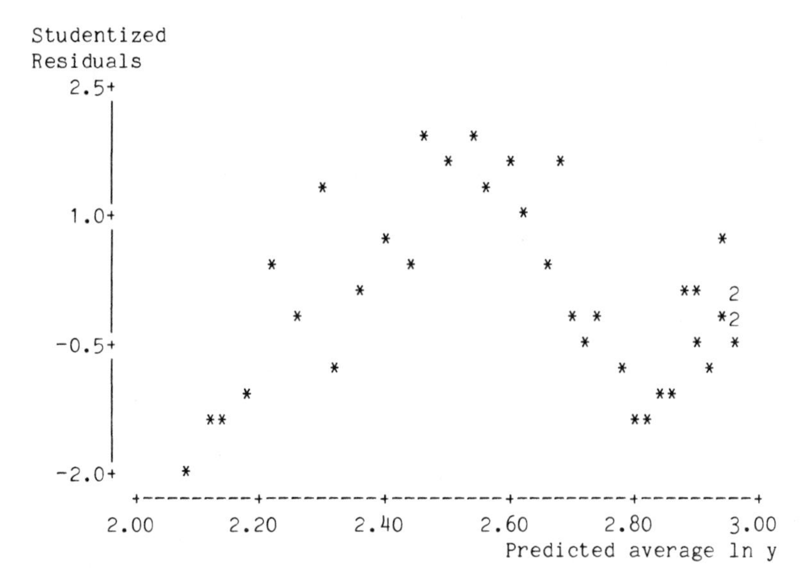

FIGURE 3.9.2 Cyclical pattern of residuals obtained by averaging responses.

```
                    ACF of the studentized residuals.
          -1.0 -0.8 -0.6 -0.4 -0.2  0.0  0.2  0.4  0.6  0.8  1.0
            +----+----+----+----+----+----+----+----+----+----+
 1   0.558                           xxxxxxxxxxxxxxx
 2   0.515                           xxxxxxxxxxxxxx
 3   0.362                           xxxxxxxxxx
 4   0.180                           xxxxx
 5   0.140                           xxxx
 6  -0.045                          xx
 7  -0.088                         xxx
 8  -0.229                     xxxxxxx
 9  -0.287                    xxxxxxxx
10  -0.420                 xxxxxxxxxxx
11  -0.505              xxxxxxxxxxxxxx
12  -0.416                 xxxxxxxxxxx
13  -0.429                xxxxxxxxxxxx
14  -0.292                    xxxxxxxx
15  -0.281                    xxxxxxxx
16  -0.180                      xxxxxx
```

FIGURE 3.9.3 ACF of residuals from the averaged responses.

PROBLEMS

3.1 Consider the following ANOVA

Sources of Variation	Degrees of Freedom	Sum of Squares
Regression	3	0.982336
Residual	9	0.017664
Total	12	1.000000

(i) Write a model for this regression.
(ii) How many observations were there in the data?
(iii) How do you explain the peculiar value for the sum of squares for total?
(iv) Evaluate R-SQUARED, the coefficient of determination.

3.2 Three predictor variables, x_1, x_2, x_3, and one dependent variable, y, are in correlation form and the correlation matrix is:

x_1	x_2	x_3	y
1	.78	.50	.92
	1	.39	.88
		1	.40

The residual degrees of freedom are 17.

(i) How many observations were there in the experiment?
(ii) For the model $y = \beta_1 x_1 + \beta_2 x_2 + \beta_3 x_3 + \varepsilon$, the R^2 value is .914. Is this significant at the 5% level?
(iii) If only one predictor variable is used, which one should it be and what would be the R^2 value for this model?
(iv) Plot Mallows' C_p statistic to decide on the best model, given these R^2 values:-

Variables	R^2
1,2	.909
1,3	.851
2,3	.770

3.3 At various times after birth, nine calves were tested for serum gamma glutamyl transpeptidase (GGT). Part of this data was used in Problem 1.4. A model was fitted

$$y = \beta_0 1 + \beta_1 x_1 + \beta_2 x_2 + \cdots + \beta_{10} x_{10} + \varepsilon$$

where $y = \log_e$(GGT in International units per litre), x_1 = no. days after birth, and x_2 = 1 or 0 depending on whether the observations refer to calf no. 1 or not. Likewise the variables x_3 - x_{10} were dummy variables for the other calves. Parts of the computer printout are shown below.

(i) The program declared that x_{10} was highly correlated with other predictor variables so that it was omitted from the equation. Explain why.
(ii) Write the prediction equation for the ninth calf.
(iii) Test H: $\beta_9 = 0$. What does this test tell us?
(iv) Can you test H: $\beta_2 = 0$. What does this test tell us?
(v) Can you test H: $\beta_8 = \beta_9 = 0$ from the information given?
(vi) The scientist who supplied the data would really prefer the simpler model

$$y = \beta_0 1 + \beta_1 x_1 + \varepsilon$$

Would you advise him to use this simpler model?
(vii) Why do you think that a logarithmic transformation was used? Does it seem to be appropriate in the light of the standard residuals for calf no. 1 (the standard residuals for the other calves follow a similar pattern).

Computer Printout

The regression equation is

$$y = 6.06 - 0.102\,x_1 + 0.515\,x_2 - 0.144\,x_3 + 0.344\,x_4 + 0.0063\,x_5 - 0.119\,x_6 + 0.476\,x_7 + 0.263\,x_8 - 0.124\,x_9$$

ANOVA

Due to	d.f.	SS	MS = SS/d.f.
Regression	9	130.175	14.464
Residual	94	5.996	0.064
Total	103	136.172	

Further ANOVA

SS explained by each variable when entered in the order given.

Due to	d.f.	SS
x_1	1	123.884
x_2	1	1.440
x_3	1	0.795
x_4	1	0.633
x_5	1	0.102
x_6	1	0.716
x_7	1	1.747
x_8	1	0.770
x_9	1	0.083

The studentized residuals for calf no.1 are:

x_1	Studentized residuals
2	1.63
4	0.40
7	-0.55
10	-0.42
14	-0.79
17	-1.11
21	-0.78
24	0.81
35	0.83

If $(X^TX)^{-1} = \{c_{ij}\}$ the diagonal elements, c_{ii}, are 0.110, 0, 0.195, 0.167, 0.161, 0.168, 0.175, 0.161, 0.175, 0.175

3.4 Aggregating Lactation Curves
This problem follows on from Section 3.9 and involves the data of Appendix C3. The estimated coefficients in regressing ln y (logarithm of the milk yield) for a given cow against w and ln w (w being the week number) are as follows:-

Cow no.	Constant	w	ln w	R-Squared
1	2.68	-0.0376	0.174	92.2
2	3.11	-0.0385	0.244	88.2
3	3.29	-0.0192	0.0004	88.9
4	2.94	-0.0438	0.257	94.8
5	1.56	-0.0839	0.771	85.6
av ln y	2.72	-0.0446	0.289	94.8

For this last model we have averaged the data for the five cows.

(i) Write the prediction equation in terms of y for cow no.1.

(ii) Average the coefficients for the five cows, and use this prediction equation to find the predicted milk yields and residuals for cow no.1. How well does this model fit?

(iii) Comment on the average values of the coefficients you obtained in part (ii) and the coefficients from the regression of the average values of ln y shown above. In this example, is it reasonable to aggregate the data for the five cows?

4

PECULIARITIES OF OBSERVATIONS

4.1 INTRODUCTION

In Chapter 3 we considered the relationship between variables **y** and $X = (\mathbf{x}_1, \mathbf{x}_2, \ldots, \mathbf{x}_k)$, that is, the relationship between the column vectors. In this chapter, we turn our attention to the rows or individual data points

$$(x_{1i}, x_{2i}, \ldots, x_{ki}, y_i)$$

We have already seen that the variances of the predicted values and of the residuals depend on the particular values of the predictor variables, x. Peculiar values of the x's could be termed sensitive, or high leverage, points as will be explained in Section 2. On the other hand, the observed value of y may be unusual for a given set of x values and y may then be termed an outlier as explained in Section 3.

Also, in Sections 6 through 8, the emphasis is again on the variables in the model and perhaps these topics, more logically, should fall into Chapter 3. They have been added for completeness as they are topics often referred to in other texts.

4.2 SENSITIVE, OR HIGH LEVERAGE, POINTS

All x values are not equal, at least in the influence they have on a least squares prediction curve. Consider the very simple case of a univariate prediction line through the origin where the x values are x = 1, 2, 3, 4, 10. Because the method of least squares minimizes the sum of squares of the residuals the prediction could either be (i) or (ii) in Figure 4.2.1 depending on whether the observed y value is at the point marked "a" or "b". x = 10 is called a <u>high leverage point</u> as it has an unduly large effect on the prediction line. As n is 5, the 5×5 projection matrix is

$$P = \mathbf{x}(\mathbf{x}^T\mathbf{x})^{-1}\mathbf{x}^T = \begin{vmatrix} 1 \\ 2 \\ 3 \\ 4 \\ 10 \end{vmatrix} (130)^{-1} (1, 2, 3, 4, 10)$$

Some properties of the projection matrix $P = \{p_{ij}\}$ are as follows (many have already been discussed):

(i) Predicted y is $\hat{\mathbf{y}} = P\mathbf{y}$. For example

$$\begin{aligned} \hat{y}_1 &= p_{11}\, y_1 + p_{12}\, y_2 + p_{13}\, y_3 + p_{14}\, y_4 + p_{15}\, y_5 \\ &= (y_1 + 2y_2 + 3y_3 + 4y_4 + 10y_5)/130 \end{aligned}$$

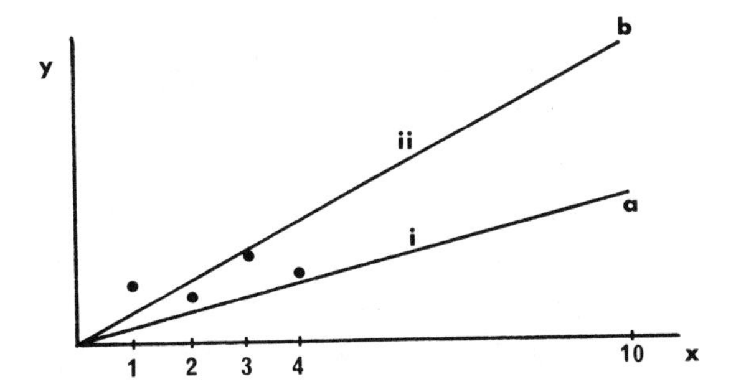

FIGURE 4.2.1 Presence of a high leverage point.

Each observed value of y contributes to the predicted value and surprisingly y_5 has 10 times as much influence on $\hat{y}_1$ as does y_1. In general, we could say that p_{ij} is the effect on $\hat{y}_i$ of y_j.

(ii) The variance-covariance matrix of $\hat{\mathbf{y}}$ is $P\sigma^2$. That is,

$$\text{var } \hat{y}_i = p_{ii}\sigma^2 \text{ and } \text{cov}(\hat{y}_i, \hat{y}_j) = p_{ij}\sigma^2 \qquad (4.2.2)$$

In particular, for our simple example var $\hat{y}_5$ = 100/130 σ^2 which is far higher than for any other $\hat{y}_i$. This confirms an expectation that $\hat{y}_5$ will vary considerably more than any other predicted value. As we expect, some covariances, particularly cov($\hat{y}_1$, $\hat{y}_5$), are quite large.

(iii) As P is symmetrical and idempotent, it can be shown that

$$0 \leqq p_{ii} \leqq 1 \quad \text{and} \quad \Sigma\, p_{ii} = k,$$

where k is the rank of the matrix. The average value of p_{ii} is then k/n.

(iv) Large values of p alert us to possible peculiar values of x. As a rule of thumb, we consider the i-th point a sensitive one if p>2 k/n, that is, twice as large as the average value. In the above simple example, k = 1, n = 5 so that there is one sensitive or high leverage point, namely the fifth point where p = 100/130 = 0.77, which is larger than 2 k/n = 0.4.

4.3 OUTLIERS

Both the predictor and dependent variables will have their parts to play in deciding whether an observation is unusual. The predictor variables determine whether a point has high leverage. The value of the dependent variable, y, for a given set of x values will determine whether the point is an outlier.

First, we consider the residuals, or preferably the studentized residuals. As explained in Section 1.3, it is good practice to plot the residuals against the predicted values of y and the predictor variables which are already in the model or which are being considered for inclusion in the model. A studentized residual of large absolute value may suggest that an error of measurement or of coding or some such has occurred in the response variable. If the sample size is reasonably large, the observation could be deleted from the analysis. It is always worthwhile, though, to consider such outliers very carefully for they may suggest conditions under which the model is not valid.

It should be noted that the size of the residuals depends on the model which is fitted. If more, or different, predictor variables are included in the model then it is likely that different points will show up as being potential outliers. It is not possible, then, to completely divorce the detection of outliers from the search for the best model.

Outliers may also be obscured by the presence of points of high leverage for these tend to constrain the prediction curve to pass close to their associated y values. These interrelated effects should warn us to tread cautiously as there is no guaranteed failsafe approach to the problem. Many solutions have been suggested and the interested reader may consult Hoaglin and Welsch (1978) . We shall not consider the tests in detail which are contained in this article. To decide whether the i-th observation is an outlier, a fruitful approach is to see the effect that would result from the omission of the i-th row of the data. In particular, how would this omission affect the residual at the point and how would it affect the slope of the prediction line?

4.4 WEIGHTED LEAST SQUARES

One of the assumptions of least squares is that the variance remains constant for all values of the predictor variables. If it is thought that this assumption is not valid it may be possible to modify the least squares method by giving different weights to different observations. We shall illustrate this method of weighted least squares in the following example.

Example 4.4.1 Wild Deer

The New Zealand Forest Service monitors the shooting of wild deer. For the years 1977-79, there were 712 male deer shot in the Ruahine ranges in the North Island. One of the measurements of the deer recorded was the jaw length and another was the age. The data is

TABLE 4.4.1 Age and Jaw Length of Deer

Age	Age (midpoint of category)	Number of animals	Mean jaw length	Standard deviation of jaw length
< 1	0.5	71	184.9	21.24
1+	1.5	250	231.8	16.64
2+	2.5	210	254.8	15.30
3+	3.5	59	272.5	11.08
4+	4.5	44	271.5	12.93
5+	5.5	34	278.3	10.03
6+	6.5	12	284.9	7.24
7+	7.5	9	281.8	11.05
8+	8.5	8	285.0	13.30
9+	9.5	7	272.3	17.73
≥10	?	8	. . .	. . .

shown in Table 4.4.1, and the aim of this exercise is to model the jaw length against age. The last age category was omitted from calculations as it is difficult to decide on its mid-point.

From the plot of mean jaw length against age in Figure 4.4.1 it appeared that it would be advisable to transform the age to its natural logarithm and, when this was done, the prediction equation was

$$\text{mean jaw length} = 218 + 32.8 \ln \text{age}$$
$$\text{or} \quad y = 218 + 32.8 \quad x$$

The R^2 value was 0.91 which is quite high and, in fact, is misleading as the R^2 value will tend to be high when the dependent variable is the mean over a number of observations. As the numbers of animals within age categories vary so markedly we may feel that more weight should be given to the means of the age categories with large numbers of deer and less weight to the smaller categories. As it turns out, a method of weighted least squares will follow this procedure and at the same time satisfy the requirement of constant variance. Let

$$y_i = \beta_0 + \beta_1 x_i + \varepsilon_i \tag{4.4.1}$$

If we assume that the variances of the observations is the same in

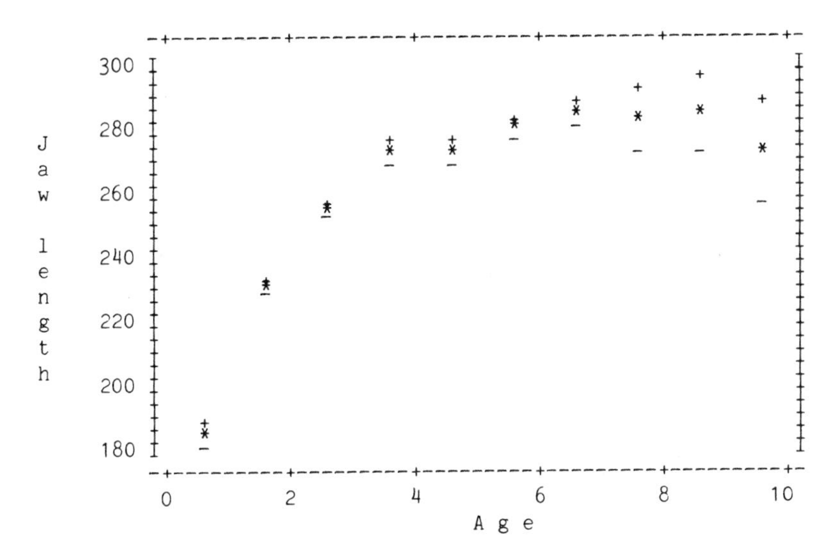

FIGURE 4.4.1 Plot of mean jaw length (*) against age, with 95% confidence intervals (from - to +).

each group, then, as the y_i values are means over n_i observations

$$\text{var } y_i = \text{var } \varepsilon_i = \sigma^2/n_i$$

To obtain constant variance, we multiply through (4.4.1) by $\sqrt{n_i}$ to give

$$\begin{aligned} \sqrt{n_i}y_i &= \sqrt{n_i}\beta_0 + \sqrt{n_i}\beta_1 x_i + \sqrt{n_i}\varepsilon_i \\ t_i &= \alpha_0 + \beta_1 z_i + \delta_i \end{aligned} \tag{4.4.2}$$

Notice that var t_i = var δ_i = σ^2 so that the least squares estimate of β_1 from (4.4.2) is found by regressing t_i on z_i to give

$$\begin{aligned} b_w &= \Sigma(z_i - \bar{z})(t_i - \bar{t})/\Sigma(z_i - \bar{z})^2 \\ &= \Sigma(x_i - \bar{x}_w)w_i(y_i - \bar{y}_w)/\Sigma w_i(x_i - \bar{x}_w)^2 \end{aligned} \tag{4.4.3}$$

In this weighted estimator, the weights, (the w's) are inversely proportional to the variance, and the variance is inversely proportional to the sample size. Therefore we have

$$w_i = k\, n_i$$

Notice that the actual size of the constant k is irrelevant as it is only necessary to know the relative sizes of the variances, and hence the weights. The quantity $\bar{x}_w = \Sigma\, n_i x_i / \Sigma\, n_i$ could be called the weighted mean of x. If the model (4.4.1) is written in vector form we have

$$\mathbf{y} = X\boldsymbol{\beta} + \boldsymbol{\varepsilon} \quad \text{with var } \boldsymbol{\varepsilon} = V\sigma^2 \qquad (4.4.4)$$

where $V = \text{diag}\{1/n_i\}$.

The weighted least squares estimator is

$$\mathbf{b}_w = (X^T W X)^{-1} X^T W \mathbf{y} \qquad (4.4.5)$$

Here $W = V^{-1}$. The prediction curve is now

$$\text{mean jaw length} = 216 + 36.8 \ln \text{age}$$

In this case, the weighted least squares estimates are very little different from the ordinary least squares estimators.

We may question whether the variances of the observations are the same for each group as the variances in the sample vary from category to category. We could estimate the variance of the deviation term in (4.4.4) by

$$\widehat{\text{var}}\ \boldsymbol{\varepsilon} = \hat{V}, \quad \text{where } \hat{V} = \text{diag}\ \{s_i^2 / n_i\}$$

In (4.4.5) we could estimate W by $\hat{V}^{-1}$ so that

$$\hat{W} = \text{diag}\ \{n_i / s_i^2\}$$

This gives a prediction curve

$$\text{mean jaw length} = 218 + 36.8 \ln \text{age}$$

If the model (4.4.4) is indeed the true one, the weighted estimator is unbiased as

$$E(\mathbf{b}_W) = (X^T W X)^{-1} X^T W (X \boldsymbol{\beta}) = \boldsymbol{\beta}$$

Although it is not easy to prove, the weighted least squares estimator will have smaller variance than the ordinary least squares estimator, that is

$$\operatorname{var} \mathbf{b}_W \leqq \operatorname{var} \mathbf{b}$$

Of course, the properties do not strictly hold if W is not known and is replaced by an estimate, but we can assume that the properties will be approximately true if the estimate is a good one. We can form an ANOVA table using

$$SST = \mathbf{y}^T W \mathbf{y} \quad \text{and}$$

$$SSR = \mathbf{y}^T W X (X^T W X)^{-1} X^T W \mathbf{y} .$$

4.5 MORE ON TRANSFORMATIONS

We list the assumptions we make about the linear model,

$$\text{response} = \text{mean} + \text{deviation} \tag{4.5.1}$$

(i) An additive model. (The terms on the right are added not multiplied.)
(ii) The mean has a reasonably simple structure.
(iii) An additive error structure.
(iv) The deviations (and the responses) have constant variance.
(v) The deviations (and the responses) are normally distributed.

Sometimes a transformation is needed to make these assumptions believable. Of course, it is possible that a transformation may achieve one, but not all, of these aims. On the other hand, some authors suggest that a single transformation often brings multiple blessings. We consider each of these five assumptions in turn.

(i) An additive model. This is the linear part of linear models. Some models, such as (4.5.2), are nonlinear and we do not consider their solutions in this book. Some nonlinear models can be linearized by a transformation such as (4.5.3) and (4.5.4). An intrinsically nonlinear model is

$$y = \mu(x_1^{\alpha} + x_2^{\beta}) + \varepsilon \quad (4.5.2)$$

On the other hand (4.5.3) becomes a linear model by taking logarithms to any base as shown next.

$$y = \mu x_1^{\alpha} x_2^{\beta} \varepsilon \quad (4.5.3)$$

$$\log y = \log \mu + \alpha \log x_1 + \beta \log x_2 + \log \varepsilon$$

A reciprocal transformation changes the nonlinear

response = reciprocal(mean + deviation) (4.5.4)

to the linear

reciprocal(response) = mean + deviation.

(ii) The mean has a reasonably simple structure. This is not necessary for the statistical properties of the model but it is helpful in explaining results. In the deer example of the previous section, it would have been possible to use a polynomial in the age of the animal, but the coefficients of a polynomial, particularly of the higher powers, rarely have a simple practical interpretation. By using the logarithm of the age some complexity is avoided while still giving a model which fits the data well.

(iii) The error structure is additive. In (4.5.3), the error structure is multiplicative as the mean is multiplied by the deviation. By taking logarithms, the error term is added to the mean.

(iv) The deviations (and the responses) have constant variance. This ensures that the coefficient estimates have small variance. Transformations may be suggested by theoretical or practical considerations. An early paper on transformations was by Bartlett (1947) who derived a formula for obtaining a transformation. He argued that if the variance of the response, y, could be assumed to be a function of its mean, namely variance = f (mean), then a transformation g(y) is needed to give constant variance. Bartlett showed that

$$g = \int(1/\sqrt{f}) \quad (4.5.5)$$

For example, if the y values are count data (number of plants in a given area, number of worms in a certain volume etc) then, under certain restrictions, y may follow a Poisson distribution. For this distribution the variance equals the mean, and (4.5.5) suggests that the appropriate transformation is the square root. Bartlett also suggested the use of $\sqrt{(y+1/2)}$ when small numbers, especially zeros, are involved.

If the y values follow a binomial distribution, $g(y) = \sin^{-1}\sqrt{y}$ expressed in radians. On the other hand, if the variance of y increases with the square of the mean, a logarithm transformation is called for. Even if there is no theoretical basis for a transformation, a consideration of residual plots may suggest one which would lead to an improvement in the fit.

(v) The deviations (and the responses) are normally distributed. If the sample size is not too small, then tests on the residuals could be conducted to see if they can be assumed to follow a normal distribution.

4.6 EIGENVALUES AND PRINCIPAL COMPONENTS

With all linear models, the ideal situation occurs when the predictor variables are orthogonal. More specifically, we would like them to be orthogonal when they are in the form of deviations from their respective means as this ensures that they are uncorrelated. Along with the assumption of normality of the dependent variable this leads to the independence of the estimates of the coefficients and, indeed, these become equal to their value if the dependent variable was regressed separately on each predictor variable.

At the other extreme, high correlations among the predictor variables can lead to problems both of a computational and statistical nature. High correlations lead to difficulties in obtaining the inverse of $S = X^T X$ for it could be close to being singular, the determinant being close to zero. Numerical methods have been developed to reduce such difficulties but there remains statistical problems of interpreting the estimates. We could not be sure whether a variable was having a strong influence on y directly or whether the influence was due to other variables which are highly correlated with it. In fact, the estimates **b** may have little meaning at all as

$$\text{var}\ \mathbf{b} = S^{-1}\sigma^2$$

and this variance could be large if high correlations were causing S to be nearly singular. This problem is often termed collinearity or multicollinearity and a number of means have been sought to reduce its ill-effects. We shall have more to say about this at the end of this section and in Section 4.7.

In this section, we describe how to transform the predictor variables to principal components. In Section 3.4, we considered one method of obtaining orthogonal predictor variables by adding them sequentially to the model. The method of principal components also produces orthogonal predictor variables but has the advantage that it highlights the structure of the X matrix. We consider how in a simple example.

In the scatter diagram, Figure 4.6.1, the two variables are scaled to be in correlation form. Suppose that their correlation is 0.8. The matrix $X = (\mathbf{x}_1 \ \mathbf{x}_2)$ and has dimensions $n \times 2$.

$$X^T X = \begin{vmatrix} \mathbf{x}_1^T \mathbf{x}_1 & \mathbf{x}_1^T \mathbf{x}_2 \\ \mathbf{x}_2^T \mathbf{x}_1 & \mathbf{x}_2^T \mathbf{x}_2 \end{vmatrix} = \begin{vmatrix} 1.0 & 0.8 \\ 0.8 & 1.0 \end{vmatrix}$$

If the axes are now rotated so that the observations are expressed in terms of w_1 and w_2 some of the structure of the data is more clearly revealed. The points fall close to the line, the w_1-axis, and the points are spread out more along this axis than the w_2-axis.

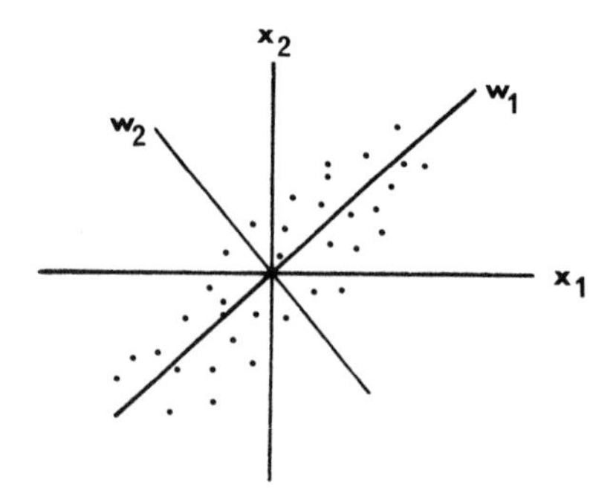

FIGURE 4.6.1 Scatter diagram in terms of x_1, x_2 and rotated axes w_1, w_2.

We may guess correctly that we can estimate the coefficient associated with w_1 much more precisely than that associated with w_2.

In this simple example,

$$X = (\mathbf{x}_1 \quad \mathbf{x}_2) \text{ and has dimensions } n\times 2.$$

X^TX can be diagonalized by the 2×2 matrix $C = (\mathbf{c}_1, \mathbf{c}_2)$ where $\mathbf{c}_1 = (1 \quad 1)^T/2$ and $\mathbf{c}_2 = (1 \quad -1)^T/2$. That is, $C^T(X^TX)C = W^TW = \Lambda$ which is a diagonal matrix with diagonal elements $\lambda_1 = 1.8$ and $\lambda_2 = 0.2$. The matrix C has been chosen to be orthonormal ($C^TC = I$, the identity matrix) with the larger element of Λ having the subscript 1.

It is always possible to find a matrix C to diagonalize a symmetric matrix and C is unique if it is given the above properties. The $\mathbf{c}_i$ are called eigenvectors or latent vectors, the λ_i are called eigenvalues or latent values and the $\mathbf{w}_i$ are called principal components. Let

$$\begin{aligned} \mathbf{y} &= \beta_1 \mathbf{x}_1 + \beta_2 \mathbf{x}_2 + \varepsilon &&= \alpha_1 \mathbf{w}_1 + \alpha_2 \mathbf{w}_2 + \boldsymbol{\varepsilon} \\ &= X\boldsymbol{\beta} + \boldsymbol{\varepsilon} &&= W\boldsymbol{\alpha} + \boldsymbol{\varepsilon} \end{aligned} \quad (4.6.2)$$

The projection matrix onto W can easily be shown to be equal to the projection matrix onto X. Thus the predicted y's and the MSE will be the same for each model. There will be some advantage, though, in using the second form of the model which stems from the fact that the principal components are orthogonal. With k predictors

(i) The sum of squares for regression can be split into k orthogonal parts which means that the F statistics to test each hypothesis, H: $\alpha_i = 0$, will be independent of each other.

(ii) $$\text{var } \mathbf{a} = (W^T W)^{-1} \sigma^2 = \Lambda^{-1}\sigma^2$$

$$\text{or var } a_i = \sigma^2/\lambda_i \quad \text{for } i = 1,2, \ldots ,k$$

Large eigenvalues lead to a's with small variance, or saying this in another way:-

We can estimate coefficients more efficiently in the directions of the principal component associated with the large eigenvalues.

The converse is that if an eigenvalue is very small then the estimate of the coefficient has such large variance that it is of limited interest.

(iii) If some of the correlations between predictor variables are relatively large, it will turn out that the determinant

$$\det(X^T X) = \det(W^T W) = \text{product of the } \lambda_i$$

will be close to zero as at least one of the λ's will be close to zero.

(iv) To remove collinearity we could omit the principal component corresponding to the smallest eigenvalue. However, the correlation between y and this component may be large. It can be seen that the sum of squares for regression for the i-th principal component is

$$SSR_i = \lambda_i a_i^2 = r_{yi}^2/\lambda_i \qquad (4.6.3)$$

We could follow the rule of removing this principal component if its eigenvalue is small compared with other eigenvalues or if its sum of squares is not significant.

4.7 RIDGE REGRESSION

We have seen that high correlations among some predictor variables have adverse effect on the least squares coefficient estimates and that they may fall outside of a range or have a different sign than would be expected from past experience. Small changes in the dependent variable could result in large changes in the estimates. Another way of saying this is that the variances of some estimates may be large as

$$\text{var } \mathbf{b} = S^{-1}\sigma^2 \qquad \text{where } S = X^T X$$

$$\text{and var } b_i = c_{ii}\sigma^2 \qquad \text{where } S^{-1} = C = \{c_{ij}\}$$

These diagonal elements, the c's, have been termed <u>variance inflation factors</u> (VIF) and some of those will often be very large in the presence of high correlations. To cope with this problem of multicollinearity, A E Hoerl and R W Kennard in 1970 suggested that these variance inflation factors would be reduced and the coefficient estimators stabilized by using the following estimator which they called the ridge, or ridge regression, estimator given by

$$\mathbf{b}_R = (S + kI)^{-1} X^T \mathbf{y} \tag{4.7.1}$$

Clearly, if k were very large the kI term would dominate S, the matrix of sums of squares and cross products, and the variance inflation factors defined as the diagonal elements of

$$C = (S + kI)^{-1} \tag{4.7.2}$$

would be reduced, hence stabilizing the coefficient estimates.

Under the usual conditions of the linear model, the ridge estimator is biased and the bias increases with the size of k. Proponents of this method suggest, though, that in practice a small value of k can be chosen, usually less than 0.2, which greatly reduces the variability of the estimates while only resulting in small bias. One way to choose the approximate value of k is to plot the values of the estimates and the VIF against k. These plots are called ridge traces, and from them the smallest value of k which stabilizes the coefficients can be chosen. This procedure is illustrated in the following example.

Example 4.7.1 Chicken feed mixture

G S Wewala (1980) analyzed a mixture experiment in which chickens were fed a mixture of maize, fishmeal and soy bean and their weight noted when 1000 days old. At each of ten design points, 32 chickens were reared giving the results in Table 4.7.1. As proportions add

TABLE 4.7.1 Chicken Weight Gains Under Different Diets

Design points	Proportion of			Mean final weight
	Maize, x_1	Fishmeal, x_2	Soy bean, x_3	
T_1	0.8	0.2	0.0	0.4473
T_2	0.7	0.3	0.0	0.4701
T_3	0.3	0.3	0.4	0.4532
T_4	0.3	0.2	0.5	0.4729
T_5	0.5	0.0	0.5	0.4690
T_6	0.8	0.0	0.2	0.4143
T_7	0.8	0.1	0.1	0.4389
T_8	0.65	0.15	0.2	0.4718
T_9	0.5	0.15	0.35	0.4606
T_{10}	0.5	0.3	0.2	0.4650

to one, only terms in the first two ingredients were included in the model which was a full quadratic in these two ingredients, namely

$$y = \beta_1 x_1 + \beta_2 x_2 + \beta_3 x_1^2 + \beta_4 x_1 x_2 + \beta_5 x_2^2 + \varepsilon \quad (4.7.3)$$

The ridge traces of the coefficient estimates and the variance inflation factors are shown in Figures 4.7.1 and 4.7.2. For small values of k, the estimates change in relation to each other but seem to settle down from k equals 0.004 onwards. With the possible exceptions of the coefficients b_1 and b_3 the estimates decrease in size in a monotone fashion as k increases. This trend would continue for even larger values of k.

The variance inflation factors also decrease in size quite dramatically as k increases. What value of k should be chosen? It would

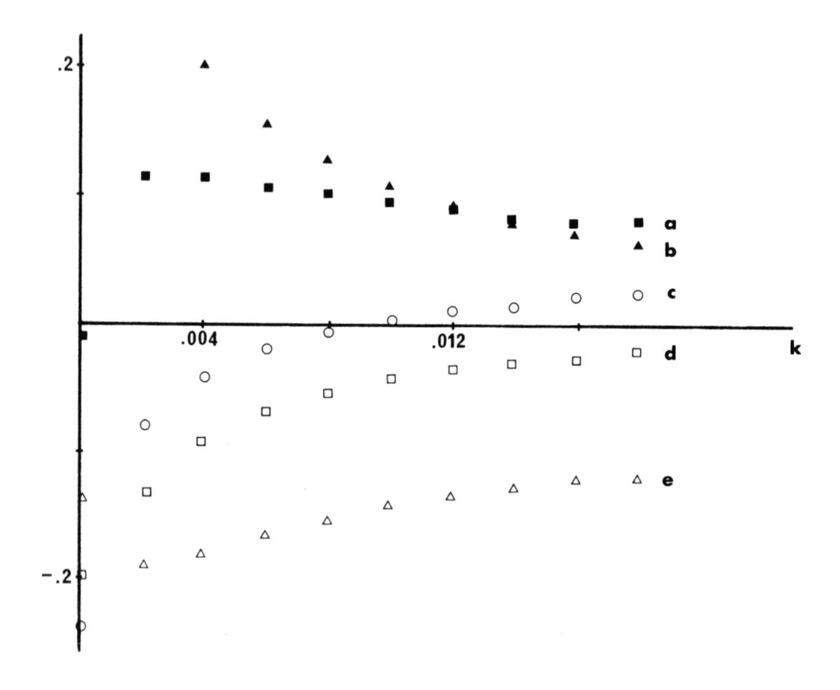

FIGURE 4.7.1 Ridge trace for coefficients for the chicken feed data. Note: a, b, c, d and e refer to the estimates for β_1, β_4, β_2, β_5, and β_3 respectively.

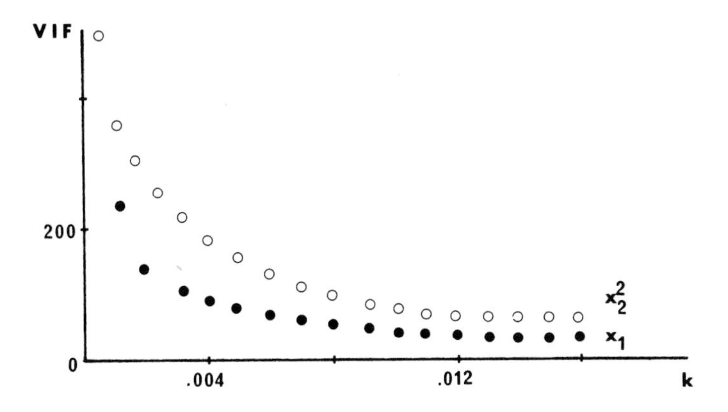

FIGURE 4.7.2 Variance inflation factors of two coefficients for the chicken feed data.

appear that k = 0.008 (say) is quite small so that the introduced bias would be negligible but the gains, in terms of stabilizing the coefficients and their variances, are quite impressive.

The ridge trace, however, can be misleading. One of the problems is that of finding the appropriate transformation of the variables.

In the above example, if the variables are written in correlation form, the ridge trace does not exhibit the same initial fluctuations but each estimate monotonically decreases in size as k increases and the variance inflation factors commence, with k = 0, at a much smaller value. It is usually recommended that variables should be written in correlation form and the value of k should be chosen so that the largest VIF is no more than one hundred times the smallest VIF.

Ridge regression has attracted much attention in recent years. In reviewing the advances and criticisms of their estimator ten years later, Hoerl and Kennard (1981) summarized over 240 articles which had been published in reputable journals. Some of the main areas of research have been

(i) Developing a mathematical expression for k rather than relying on a visual estimate. The data itself is used to obtain an estimate to optimize a particular property such as mean square error. Unfortunately, in this case the distribution of the ridge estimator is hard to ascertain and it is not all that obvious that the ridge estimator is better than least squares except under certain conditions.

(ii) Finding different values of k for different coefficients. In the extreme, a matrix K is postulated instead of the single value of k.

(iii) Using prior knowledge to shrink the estimator toward a particular value other than zero. From the ridge trace of the estimated coefficients in Figure 4.7.1, it is clear that each estimate shrinks to zero as k increases. If one coefficient is thought to be close to be 2 (as dictated by theory, prejudice or past experience) then it could be argued that this coefficient should be shrunk towards this value rather than towards zero.

4.8 PRIOR INFORMATION

Quite often we know something about the vector of coefficients. For example, the j-th parameter might have to fall in the range (0, 1) for a theory to make sense. A Bayesian approach would seem appropriate. One way to include this prior knowledge, or belief, would be to assume that

$$\beta_j \sim N(0.5,\ 0.04)$$

The density function of the j-th parameter would be as in Figure 4.8.1. Note that it lies in (0,1) with high probability. Consider then the model

$$\mathbf{y} = X\boldsymbol{\beta} + \boldsymbol{\varepsilon} \quad \text{with} \quad \boldsymbol{\varepsilon} \sim N(0,\ \sigma^2 I) \tag{4.8.1}$$

and X is an n×p matrix. Suppose that the prior information about the

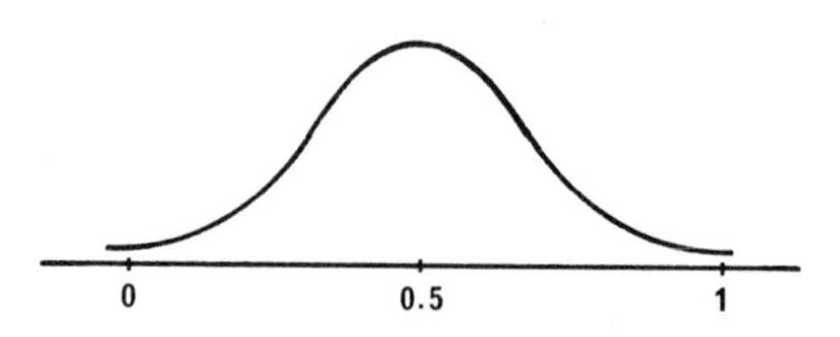

FIGURE 4.8.1 Prior distribution of β_j.

$\boldsymbol{\beta}$ vector can be written as

$$\mathbf{r} = R\,\boldsymbol{\beta} + \boldsymbol{\delta} \quad \text{with} \quad \boldsymbol{\delta} \sim N(0, \lambda^2 I) \tag{4.8.2}$$

and R is a q×p matrix. That is, we have placed q constraints on the model. We can join the two equations above into a single model.

$$\begin{vmatrix} \mathbf{y} \\ \mathbf{r} \end{vmatrix} = \begin{vmatrix} X \\ R \end{vmatrix} \boldsymbol{\beta} + \begin{vmatrix} \varepsilon \\ \boldsymbol{\delta} \end{vmatrix} \tag{4.8.3}$$

The weighted least squares estimator for β is given by

$$\mathbf{b} = \left(X^T X / \sigma^2 + R^T R / \lambda^2\right)^{-1}\left(X^T y / \sigma^2 + R^T r / \lambda^2\right) \tag{4.8.4}$$

Notice that under certain conditions $\mathbf{b}_R$ could reduce to the ridge estimator of (4.7.1), namely if

$$R^T r = 0, \quad p = q, \quad R^T R = I, \quad k = \sigma^2/\lambda^2$$

The ridge estimator can be thought of, then, as a special case of assuming prior knowledge about the coefficient vector.

It is also illuminating to indicate as we have done in (4.8.4) the competing influences of D (data) and PK (prior knowledge). Of course, $\mathbf{b}_r$ of (4.8.4) is not strictly an estimator because it involves unknown parameters of σ^2 and λ^2. These would have to be estimated to change (4.8.4) into an estimator. This would be easily possible and would indicate how prior knowledge could be included along with the current data.

4.9 CLEANING UP DATA

It should be clear from the discussion so far that unusual values of both predictor and dependent variables can have an unduly large effect on the estimates of a fitted equation. The size of this effect is sometimes staggering. As an example of this, we could consider the house price data of Appendix C5. When the price obtained for houses was regressed against the two government valuations, for house and land, the R-SQUARED value was only 0.8%. A quick scan of the data showed that there was one coding error - the price of the 43rd house was given as $840,000 instead of the correct value of $40,000. When this correction was made, the R-SQUARED value increased to 83.9 %. A 21-fold increase in one of the prices resulted in a decrease of 100-fold in the R-SQUARED value. The correlations of the price with the other variable were likewise affected by this one coding error and these are shown below with the incorrect values in parenthesis. The mistake in the price was quite obvious in this example but perhaps there are other smaller errors. For example, the area of the section for the first house is much larger than for any other house. If the correct value is 0.07 instead of 1.07 hectares, the correlations of SECTION with the other variables are affected dramatically as shown in Table 4.9.1. On the other hand, this one value of the predictor variable does not affect the R-SQUARED value very much in the model fitting the price to the other four variables.

It pays to approach every data set with some skepticism and attempt to search out discrepancies. Quite often data is handled by many people from the original measurement stage to coding and scaling and there are possibilities for errors at each step. Checks should

TABLE 4.9.1 Correlations Between Variables in the House Price Data

	Price	GV-house	GV-land	Size	Incorrect price[a]
GV-house	0.866				0.060
GV-land	0.698	0.510			0.088
Size	0.715	0.727	0.546		-0.030
Section	0.063	0.023	0.066	-0.012	-0.024
(Section	0.670	0.478	0.718	0.463)[b]	

a. This column shows the correlations with price when the 43rd value is $840,000.

b. This row shows the correlations with section when the 1st value is 0.07.

be made visually by graphing, or editing should be carried out to remove impossible or highly unlikely readings. If the data set is large it may be advisable to remove any points which are suspect before attempts are made to model the data. One should also keep in mind that unusual predictor values may be outside of the range where the model could be expected to be a good fit.

PROBLEMS

4.1 Refer to the house price data of Appendix C5.

(i) When the first entry for the size of section was changed to 0.07, the prediction equation for price on the four predictor variables was:

PRICE = 4902 + 1.37 GV-HOUSE + 0.957 GV-LAND + 327 SIZE + 142196 SECTION

giving R-SQUARED = 84.1 PERCENT, and S = 7301.

With the first entry set at 1.07, the equation was:

PRICE = 8448 + 1.43 GV-HOUSE + 1.52 GV-LAND + 348 SIZE + 3269 SECTION

giving R-SQUARED = 84.1 PERCENT
and R-SQUARED = 82.7 PERCENT, ADJUSTED FOR D.F.

(a) Explain the differences in these two fitted equations.Are the changes in the direction and of the size that you would have expected?

(b) Why is the adjusted R-SQUARED value similar to the R-SQUARED value in this case?

(ii) Part of the MINITAB printout for this model was:

ROW	GV-HOUSE	Y PRICE	PRED.Y VALUE	ST.DEV. PRED. Y	RESIDUAL	ST.RES
1	14700	48000	49799	7255	-1799	-2.18RX
4	30600	66000	82737	3143	-16737	-2.54R
17	21500	74000	57291	1879	16709	2.37R
19	35900	114000	108598	4985	5401	1.01 X
39	8600	64000	44783	2446	19217	2.79R

R DENOTES AN OBS. WITH A LARGE ST. RES.
X DENOTES AN OBS. WHOSE X VALUE GIVES IT LARGE INFLUENCE.

(a) Comment on possible reasons why these points were marked with an 'X' or 'R'.

(b) Which, if any, of these points should be removed?

(c) Select one point and show how the studentized residual, ST.RES., could be evaluated from the other information listed.

(iii) The four predictor variables were then scaled to be in correlation form and MINITAB printed out the 'X' and 'R' points. Why did they and the studentized residuals remain the same whereas the other values changed?

ROW	GV-HOUSE	Y PRICE	PRED. Y VALUE	ST.DEV. PRED. Y	RESIDUAL	ST.RES.
1	-0.057	-0.04859	-0.03378	0.05972	-0.01481	-2.18RX
4	0.229	0.09958	0.23736	0.02588	-0.13778	-2.54R
17	0.065	0.16544	0.02789	0.01547	0.13755	2.37R
19	0.324	0.49473	0.45026	0.04104	0.04447	1.01 X
39	-0.166	0.08312	-0.07508	0.02014	0.15820	2.79R

(iv) Regressions were run of price on subsets of the predictor variables,which were renamed A,B,C,D instead of GV-HOUSE, GV-LAND, SIZE and SECTION, respectively. The Sequential SS for these regressions are shown below. Mallows' C_p statistic to decide on the best model for this data. Notice that for the first model, the four variables A,B,C and D are included; for the second model B,C and D etc.

FURTHER ANALYSIS OF VARIANCE - SS EXPLAINED BY EACH VARIABLE WHEN ENTERED IN THE ORDER GIVEN

DUE TO	DF	SS
REGRESSION	4	0.84104
A	1	0.75056
B	1	0.08839
C	1	0.00137
D	1	0.00071
REGRESSION	3	0.64705
B	1	0.48658
C	1	0.15888
D	1	0.00160
REGRESSION	3	0.76820
C	1	0.51109
D	1	0.00515
A	1	0.25195
REGRESSION	3	0.83954
D	1	0.00398
A	1	0.74844
B	1	0.08712
REGRESSION	2	0.76585
A	1	0.75056
C	1	0.01529
REGESSION	2	0.48686
B	1	0.48658
D	1	0.00028

4.2 For the model $E(y) = \beta_0 + \beta_1 x_1 + \beta_2 x_2 + \beta_3 x_3$, the computer printout was

Coefficient	St Dev	T-Ratio
β_0 = -14.3	32.4	-0.44
β_1 = -0.46	1.01	-0.45
β_2 = 1.25	0.37	3.41
β_3 = 0.0089	0.0079	1.13

Due to	Analysis of variance DF	SS	MS
Regression	3	1893.6	631.2
Total	20	2069.2	

Further analysis of variance - SS explained by each variable when entered in the order given.

Due to	DF	SS
Regression	3	1893.6
x_1	1	1750.1
x_2	1	130.3
x_3	1	13.2

(i) Do the SS in the further analysis of variance suggest that the three original variables are mutually orthogonal?

(ii) Test H: $\beta_2 = 0$

(iii) If x_3 was omitted from the model, write the ANOVA.

(iv) For the model $E(y) = \beta_0 + \beta_1 x_1$, write the ANOVA and test H: $\beta_1 = 0$.

(v) If x_2 was omitted from the model, do you have enough information to calculate the ANOVA for the model $E(y) = \beta_0 + \beta_1 x_1 + \beta_3 x_3$?

4.3 Suppose that a model $y = X_1 \beta_1 + \varepsilon_1$ is fitted to some data. Suppose that the true model is really $y = X_1 \beta_1 + X_2 \beta_2 + \varepsilon$.

(i) Find the bias in $\hat{\beta}_1$ from the fitted model. Under what conditions would you expect this bias to be small?

(ii) From the fitted model show that the E(SSR) could be large even if $\beta_1 = 0$ (You may need Property 3 of Appendix B 3)

5
THE EXPERIMENTAL DESIGN MODEL

5.1 INTRODUCTION

Back in Chapter 1, Section 1, it was explained that the aim of all the work so far is to fit models to a population using data from a sample. The model so fitted shows what values of y are likely to be associated with any given values of the x's. Just as any elementary statistics text will warn that a correlation does not imply causation, so a well fitting regression model does not imply that changes in the x values will cause changes in the y variable in accordance the with model. If the model has been constructed by passive observation of sets of y's and x's, any intervention to change the x values to values which do not occur naturally will change the population to be something different from what the model describes, and so the model no longer applies.

In an experiment a population is not passively observed, but the x values are the result of intervention by the experimenter. The aim then is to intervene to choose x values in such a way as to give accurate estimates of the model parameters with a minimum of effort, and to ensure that the relationships described by the model really do describe the way changes in y are *caused* by changes in x.

5.2 WHAT MAKES AN EXPERIMENT?

One of the largest experiments ever conducted makes an interesting example, partly because it was almost a disaster. The experiment was designed to test the effectiveness of the Salk polio vaccine, and was conducted in the United States in 1954. A good account of the experiment is in Freedman, Pisani, and Purves, 1978. There were two difficulties with the experiment. First, polio is a relatively uncommon disease. The vaccine can only be tested if it is tried on a group large enough to contain a reasonable number of polio victims. For the common cold a few hundred people would be plenty, but for polio several hundred thousand were required. Second, polio is a disease which comes and goes from one season to the next. With the common cold, which most people catch most years, if the incidence drops in a test group after administering a vaccine, the drop can probably be attributed to the vaccine. But not so with polio - it may well have been a year in which the disease would have waned anyway.

The original experimental plan for the polio vaccine was to vaccinate second year school children and compare their incidence of polio with that of first and third year children. The intention is clear. The children being compared should be as alike as possible, and this plan would ensure that the geographical area and the time period would be the same for both groups. Of course they would be of different ages, but since the ages of the untreated group brackets that of the treated group one might not expect this to matter too much. A large number of experts must have considered that these remaining differences would not matter because this experiment was put into effect. 221,998 children were vaccinated, 725,173 were not, and the rate of polio following vaccination was about twice as great in the unvaccinated group as in the vaccinated group, (54 per 100,000 instead of 25).

Because many parents objected not all second year school children were vaccinated. Of these children 44 per 100,000 contracted polio, rather fewer than the unvaccinated first and third year children. It seems then that whether parents cooperated had some effect on whether their children contracted polio. There were other comparisons which were disturbing. Nine per 100,000 of the vaccinated group were falsely reported to have polio as against six per 100,000 of the control group, which suggests diagnosis differences. Of course all these figures require significance testing, but that is not the important point. The important point is that, before the experiment, the medical experts were apparently satisfied that there would be no major differences between the two groups of children, but they were wrong. The same problem afflicts all experimenters in all disciplines. How can one ensure that some unexpected factor might not invalidate the results of the experiment?

At a later stage in the polio vaccine trial an alternative experiment was proposed. For this, all first, second and third year school children were combined, and all those whose parents agreed were used. Every child received an injection, but half (selected randomly) were injected with a placebo. Neither the child nor anyone who had contact with the child knew whether the placebo or the vaccine had been used. This experiment was used on 402,974 children and the rate of polio was again about halved by the vaccine (41 per 100,000 instead of 81).

No great skill is required to see that the second experiment was much better than the first, but it is worth examining the reasons why. The crucial step in examining any experiment is to consider what differences there are between the treated group and the untreated group. For the observed control experiment a long list could be made - different ages, volunteers instead of everyone, medical treatment instead of no treatment, vaccine instead of no vaccine, for a start. Although before doing the experiment experts may have claimed that none of these differences could have caused the differences in polio incidence, much of the point of doing the experiment is lost when interpretation of the results depends heavily on expert judgment which was available without doing the experiment.

For the placebo control experiment the only differences are vaccine instead of placebo and differences introduced by the random allocation procedures. The possibility that the random allocation might have been responsible (that is by chance the susceptible children tended to fall in one group) is the possibility whose probability is measured by a significance test. Here we do not wish to become embroiled in the unresolvable debate about the place of randomization in experimentation, but we do make the claim that, at very least, using some random procedure to allocate children to groups ensures that the allocation is completely independent of any factor (known or unknown) which might possibly affect the response to the treatments. The magical quality of the toss of a coin or the book of random numbers is that there is no possible link between it and the experiment. Thus independence is guaranteed.

Randomization is necessary then, but both polio trials involved another equally important factor - replication. Here the word has a slightly different meaning from its everyday use. What it means is that each treatment is applied to more than one object. In fact, the polio trial must have been the most replicated trial ever held. Because polio is a very rare disease hundreds of thousands of children had to be treated to ensure that the difference in the numbers catching polio would be large relative to chance variation. Replication achieves two objects. By comparing different observations within the same treatment a measure of chance variability is provided, and by averaging many observations greater precision is obtained.

There is a third ingredient to any experiment, one which is not exemplified by the polio trials. In the controlled experiment the 400,000 children taking part were randomly allocated to the two treatments. It would have been possible (although not practical) to take a blood sample from each child and measure their natural immunity to polio. They could then have been divided into groups of similar immunity, and then treatments could have been applied randomly, but within each group, so that for each group exactly the same number of children would receive the inoculant and the placebo. With the very large groups involved "exactly the same" would make very little difference from random allocation, but with a more usual experiment a very worthwhile increase in accuracy can result.

Our second example is an experiment which was probably never performed, but was the example used by R. A. Fisher in his book The Design of Experiments and has since acquired the status of a classic.

A lady declares that by tasting a cup of tea made with milk she can tell whether the milk or the tea was first added to the cup. To test whether it really is this which is causing the "effect" (her judgment) the following experiment is performed. The woman is presented with eight cups of tea in random order. She is told that there are four of each kind and her task is to identify them. Fisher argued that since there are $8!/(4!4!) = 70$ ways of dividing eight objects into two groups of four, if the woman has no powers of discrimination there is one chance in 70 that her identification will be completely correct. The null hypothesis that she had no powers of discrimination would then be rejected with a a probability level of $1/70 = 1.4\%$. Note that here the whole significance level calculation is based on the fact that the experiment was randomized. A test like this is always possible, but usually involves an impractical amount of computing. In this book we will use only linear models for analysis.

Randomization is still necessary for the reasons given in the polio example, and which we will demonstrate for this example. Consider just how the experiment would be carried out. There would be eight cups in a row. Four would be randomly chosen to have milk poured into them. Then they would be filled with tea in order along the row and the milk added to the remaining four cups. The cups would then be presented to the taster in row order. Note that everything is done in row order, and that randomizing ensures that treatments are assigned independently of row order. It will not matter if the tea was gradually strengthened as it was poured out, or if there was some physiological carryover effect on the taster from one cup to the next. The randomization ensures that any extraneous effects, of which the experimenter has no knowledge, are as likely to work one way as the other, and so average out to zero if the experiment were repeated a number of times.

Of course any sensible experimenter also will try to reduce any extraneous effects as much as possible. In the tea tasting experiment one could overcome the effect of the gradually increasing strength of the tea by pouring the cups out in pairs, one with milk added first, the other with milk added last. Which one was which would still be determined randomly for each pair, but the strength of the the tea would then not confuse the issue. This is the same process as dividing the children into groups before allocating the polio vaccine or placebo.

Randomization might not always be possible. It then becomes a question of an individual's judgment whether any extraneous cause might have been responsible for observed differences, and different individuals might well judge differently.

5.2.1 Experimental Unit

We will define an experimental unit as that object or group of objects to which a treatment has been randomly assigned. Some examples will make this definition clear. An educationalist may be comparing teaching methods. Each child is given a test after being taught by the method assigned to it, so there is one observation per child. However the child is only an experimental unit if each child were individually assigned to a teaching method. If the children were first divided into small groups, and groups were assigned randomly to teaching method, then the group would become an experimental unit and the observation on it would be the average of all the observations on the individual children. If the two classes were used, each randomly assigned to a teaching method, a class would be an experimental unit, and because there would be no replication, no assessment would be possible. In an experiment comparing weedicides, various combinations of weedicides are randomly assigned to 5m×2m areas. Within each of these areas five .25m×.25m quadrats, positioned at random, are examined for percent weed cover. This is a very common type of biological experiment, and one where our definition of experimental units is important. The weedicides were assigned to the 5m×2m areas, and so these are the experimental units, not the .25m×.25m quadrats.

As a very general and important principle, remember that any treatment comparison must be based on one number per experimental unit. This number might often be a mean of several observations, but one must beware of pretending that by increasing the number of sub-samples one can increase the replication. It is experimental units which must be replicated.

5.3 THE LINEAR MODEL

The preceding comments have been far removed from the linear models of the first four chapters. For the rest of this chapter, and for all the next, we will be talking about the "completely randomized experiment". This is one which starts with one group of n experimental units which are assigned randomly to k treatments, r_i units to the i-th treatment. The data comprises n observations, y_{ij}, each of which can be classified according to treatment. The model can then be written:

$$y_{ij} = \mu + \tau_i + \varepsilon_{ij} \tag{5.3.1}$$

y_{ij} = observation on the j-th experimental unit of the i-th treatment.

$\mu + \tau_i$ = constant depending on the assigned treatment.

ε_{ij} = unexplained deviation, assumed $N(0,\sigma^2)$ and independent for each i, j.

In matrix notation this can be written

$$\mathbf{y} = X\boldsymbol{\beta} + \boldsymbol{\varepsilon} \tag{5.3.2}$$

where

$$\mathbf{y}^T = (y_{11}, y_{12}, \ldots, y_{1r_1}, y_{21}, \ldots, y_{k1}, \ldots, y_{kr_k})$$

$$\boldsymbol{\beta}^T = (\mu, \tau_1, \ldots, \tau_k)$$

$$X = \left|\begin{array}{cccccccccc} 1 & 1 & 0 & 0 & . & . & . & . & 0 \\ . & . & . & . & . & . & . & . & . \\ 1 & 1 & 0 & 0 & . & . & . & . & 0 \\ 1 & 0 & 1 & 0 & . & . & . & . & 0 \\ . & . & . & . & . & . & . & . & . \\ 1 & 0 & 1 & 0 & . & . & . & . & 0 \\ . & . & . & . & . & . & . & . & . \\ . & . & . & . & . & . & . & . & . \\ 1 & 0 & 0 & 0 & . & . & . & . & 1 \\ . & . & . & . & . & . & . & . & . \\ 1 & 0 & 0 & 0 & . & . & . & . & 1 \end{array}\right| \begin{array}{l} r_1 \text{ rows} \\ \\ \\ \\ r_2 \text{ rows} \\ \\ \\ \\ \\ r_k \text{ rows} \\ \\ \end{array} \qquad (5.3.3)$$

which now looks more like the linear model of earlier chapters, and just the same as the dummy variable models in Chapter 3, Section 8. Note that β comprises two groups of parameters, μ representing the overall mean, and the τ's representing the differences between treatments. Contrast this with the earlier regression models where each single column of the X matrix described a single factor of interest.

We must now face what appears to be a major difference between this model and the regression models. In the X matrix (5.3.3) the first column equals the sum of the remaining columns. Therefore the columns are linearly dependent (Appendix A 2.1). In fact the $n \times (k+1)$ matrix X has rank k, so that the $(k+1) \times (k+1)$ matrix X also has rank k and therefore must be singular. Its inverse does not exist and so the formulas for the projection matrix, the sum of squares and the parameter estimates cannot be used. However, geometrically, the projection operation as illustrated in Figure 5.1 of Chapter 1 is not affected, nor is the sums of squares for the ANOVA table. In the algebraic formulas the inverse can be replaced by a pseudo or generalized inverse, $(X^TX)^-$, which is defined by $(X^TX)(X^TX)^-(X^TX) = X^TX$. If X^TX is not singular, $(X^TX)^-$ will be unique and the same as the inverse. However if X^TX is singular it will not be unique. The projection matrix is $P = X(X^TX)^-X^T$ in either case. However we will not pursue this algebraic approach. The important thing is to understand the least squares geometry. Because of the patterned nature of the X matrix, estimates are simple functions of treatment means, as we shall shortly see, so there is no need to develop a special algebraic technique to find them. Nevertheless, there are more parameters than equations, and so there is no unique solution to the least squares equation. One must therefore transform to new parameters or impose restrictions on the original parameters to obtain unique solutions. There are several possibilities as we will demonstrate in the context of the following example.

Example 5.3.1 Simple data

Exptl unit	1	2	3	4	5	6	7
Treatment	A	B	B	A	C	A	C
Observation	10.1	11.3	11.0	10.7	10.8	10.5	10.9

Suppose there are three treatments A, B, and C compared on seven experimental units with the data as given in above. Using the model $y_{ij} = \mu + \tau_i + \varepsilon_{ij}$ and sorting the experimental units into treatment order gives the following equations:

$$\begin{vmatrix} 10.1 \\ 10.7 \\ 10.5 \\ 11.3 \\ 11.0 \\ 10.8 \\ 10.9 \end{vmatrix} = \begin{vmatrix} 1 & 1 & . & . \\ 1 & 1 & . & . \\ 1 & 1 & . & . \\ 1 & . & 1 & . \\ 1 & . & 1 & . \\ 1 & . & . & 1 \\ 1 & . & . & 1 \end{vmatrix} \begin{vmatrix} \mu \\ \tau_1 \\ \tau_2 \\ \tau_3 \end{vmatrix} + \varepsilon$$

$$[\text{i.e., } \mathbf{y} = X \boldsymbol{\beta} + \boldsymbol{\varepsilon}]$$

The normal equations are

$$\begin{vmatrix} 75.3 \\ 31.3 \\ 22.3 \\ 21.7 \end{vmatrix} = \begin{vmatrix} 7 & 3 & 2 & 2 \\ 3 & 3 & 0 & 0 \\ 2 & 0 & 2 & 0 \\ 2 & 0 & 0 & 2 \end{vmatrix} \begin{vmatrix} m \\ t_1 \\ t_2 \\ t_3 \end{vmatrix}$$

$$[\text{i.e., } X^T \mathbf{y} = X^T X \mathbf{b}]$$

Note that the first equation is the sum of the other equations so that to obtain a unique solution the X matrix must be changed or the parameters restricted.

(i) First, suppose that we are interested in the three treatments individually. Then an overall mean is not particularly useful, and it could be removed from the model giving

$$y_{ij} = \tau_i' + \varepsilon_{ij} \tag{5.3.4}$$

From the the normal equations the t's are the treatment means.

$$t_1' = 10.43, \quad t_2' = 11.15, \quad t_3' = 10.85$$

This is a very simple approach, but usually one is interested in whether there are differences between all the treatments, and not just in them individually. Therefore it is useful for the model to have a separate factor representing treatment differences. Later we will introduce models with several factors. Since in these cases the columns for each factor are linear dependent a more general procedure is required than just removing the overall mean.

(ii) Suppose one of the treatments (C say) is a standard with which A and B are being compared. If $\tau_3 = 0$, then τ_1 and τ_2 measure the difference between A and B and the standard. The model becomes

$$\begin{aligned} y_{ij} &= \mu' + \tau_i' + \varepsilon_{ij} \quad \text{for } i = 1,2 \\ y_{3j} &= \mu' \qquad\;\; + \varepsilon_{3j} \end{aligned} \tag{5.3.5}$$

From the normal equations

$$m' = 10.85, \quad t_1' = -0.42 \text{ and } t_2' = 0.30$$

This too is a simple approach, and can be used in any model where a group of columns representing a factor are linearly dependent. A well written computer package will detect when a column it is about to add to the model is linearly dependent on the columns already present, and leave it out. If the columns are dummy variables representing the levels of a factor the final one will be omitted automatically, resulting in parameters like the above.

(iii) Most often no one treatment is particularly special, and the treatment effect is thought of as being the difference between the treatment mean and the overall mean. This can be expressed by requiring the treatment effects to add to zero, so that there is a fifth equation to add to the normal equations:

$$t_1 + t_2 + t_3 = 0 \qquad (5.3.6)$$

The estimates are then

$$m' = (10.43 + 11.15 + 10.85)/3 = 10.81$$
(= mean of the means)

$$t_1' = 10.43 - 10.80 = -0.38$$

$$t_2' = 0.34 \quad \text{and} \quad t_3' = 0.04$$

This is the most common way of expressing treatment differences in the model.

(iv) All the procedures so far have differed from that procedure outlined in Chapter 1, Section 6. We have not considered working with deviations from the mean. If the columns corresponding to the treatment factors are expressed as deviations from their respective means they will be orthogonal to the column of 1's for μ, and μ will be estimated by the overall mean for y. For our example then, m = 75.3/7 = 10.76, which is a little different from the mean of the means in (iii). Putting 7m = 75.3 in the first normal equation gives

$$3t_1' + 2t_2' + 2t_3' = 0.0 \qquad (5.3.7)$$

and

$$t_1' = 10.43 - 10.76 = -0.33$$

$$t_2' = 0.39 \qquad t_3' = 0.09$$

In general, if $\Sigma r_i \tau_i = 0$ the columns of the X matrix representing treatments can be expressed in such a way that they are orthogonal to μ. Note that with this restriction the estimates depend on the sample sizes. This is hard to justify, because the sample sizes are usually determined by considerations which have nothing to do with the effects being measured. With survey data, where treatments are groups within a population, sample sizes might reflect the size of the groups. Then the mean value of all the y's would be the correct estimator of the population mean. However with experimental data the number of observations per treatment depend on factors (such as the precision required) which have nothing to do with the meaning of the parameters. The procedure described in (iii) therefore leads to parameters which are usually more meaningful. Of course if each treatment has the same number of observations the two procedures are the same, and we can have both orthogonality and estimates of meaningful parameters. The experiment is then said to be balanced. Unbalanced experiments always create the problem exemplified here. One can either have estimates which are independent of the sample sizes, or one can have estimates for one factor independent of the estimates for another. One cannot have both.

5.3.1 Notation

As means and totals figure prominently in experimental design results, special notation is used. If a suffix is replaced by a dot, observations are summed or averaged over the values of that suffix. For example:

$$y_{..} = \sum_{ij} y_{ij}$$

$$\bar{y}_{..} = (\sum_{ij} y_{ij})/n$$

$$\bar{y}_{i.} = (\sum_{j} y_{ij})/r_j \qquad \text{(or sometimes just } \bar{y}_i\text{)}$$

Note that $\bar{y}_{..}$ does not equal $\Sigma\, \bar{y}_i/k$ unless sample sizes are equal. Means are always of individual observations, so anything with a bar over it is of the same order of magnitude as a single observation.

5.3.2 Estimability

The estimates obtained above varied from case to case, but notice that in each case:

$$m + t_1 = 10.43, \qquad m + t_2 = 11.15, \qquad m + t_3 = 10.85$$

$$t_1 - t_2 = -0.72, \qquad t_2 - t_3 = -0.30, \qquad t_1 - t_3 = -0.42$$

For the combinations of parameters which these functions estimate the form of restriction does not affect the estimate. They are exmples of <u>estimable</u> functions, or functions for which there exist linear unbiased estimators. In symbols, a function $\mathbf{c}^T\boldsymbol{\beta}$ is estimable if a vector $\mathbf{a}$ can be found such that

$$E(\mathbf{a}^T\mathbf{y}) = \mathbf{c}^T\boldsymbol{\beta} \tag{5.3.8}$$

Now since $E(\mathbf{a}^T\mathbf{y}) = \mathbf{a}^T X\boldsymbol{\beta}$, we require $\mathbf{c} = X\mathbf{a}^T$, or $\mathbf{c}$ to lie in the space spanned by the rows of the matrix X. Notice from (5.3.3) that each of the first three rows in the X matrix correspond to the function $\mu+\tau_1$, which is the function estimated by the first of the estimates above. Also by subtracting (say) the fourth row of X from the first we obtain $\tau_1-\tau_2$, and the other functions can be obtained in a similar fashion.

$\mathbf{c}$ is a k+1 dimensional vector, so that if the k+1 columns of X are linearly independent (as they usually are in regression) the rows will span a k+1 dimensional space, any $\mathbf{c}$ will lie in that space, and any function $\mathbf{c}^T\boldsymbol{\beta}$ will be estimable. In experimental design models, the columns of X are not linearly independent, so that the space spanned by the rows of X is of dimension less than k+1, not every $\mathbf{c}$ lies in that space and not every function $\mathbf{c}^T\boldsymbol{\beta}$ can be estimated.

5.4 TESTS OF HYPOTHESIS

The first question asked of an experiment will be, is there a treatment effect? In terms of the model, are the parameters which represent treatment differences zero? This is an hypothesis about τ's and is tested by comparing the EMS's of the two models (see Chapter 2, Section 9):

$$y = \mu + \varepsilon \quad \text{and}$$
$$y = \mu + \tau_i + \varepsilon \tag{5.4.1}$$

Using the estimates derived in Chapter 2, Section 5, SSR can be written down and the F-statistic calculated without any further theory. However, we must look briefly at the form of the various sums of squares. From the geometry

$$SSR = \hat{\mathbf{y}}^T \hat{\mathbf{y}} = \mathbf{b}^T X^T X \mathbf{b} \tag{5.4.2}$$

$$= \mathbf{b}^T X^T \mathbf{y} \text{ from the normal equations}$$

and this can be calculated and will have the same value however **b** was obtained. Since the hypothesis about the treatment effect is to be tested independently of the mean we will restrict the τ's as in (5.3.7) so that they can be estimated independently of μ. Then

$$X^T \mathbf{y} = \begin{vmatrix} y_{..} \\ y_{1.} \\ y_{2.} \\ y_{3.} \end{vmatrix} \text{ which comprises totals for each parameter.} \tag{5.4.3}$$

$$\mathbf{b}^T X^T \mathbf{y} = n\, \bar{y}_{..}^2 + \sum r_i (\bar{y}_i - \bar{y}_{..})^2$$

$$(= n\, m^2 + \sum r_i\, t_i^2 \text{ from 5.3.7)} \tag{5.4.4}$$

Because the column for m is orthogonal to the columns for the t's, and because $n y_{..}^2$ is already known to be the sum of squares for a model which includes only the mean, $\Sigma r_i t_i^2$ must be <u>the</u> sum of squares for treatments. We will use the notation SS( ) for the sum of squares of a factor, where the corresponding parameter is within the brackets. The analysis of variance table is given in Table 5.1.1. The F-statistic, $MS(\tau)/s^2$, which has k-1 and n-k degrees of freedom, can be used to test the hypothesis that there is no difference between treatments. To give an example of two orthogonal factors we

TABLE 5.1.1 Analysis of Variance Table

Source	Sums of squares	d.f.	Mean sum of squares	F
Treatment	$SS(\tau) = \sum r_i (\bar{y}_i - \bar{y})^2$	k-1	$MS(\tau)$	$MS(\tau)/s^2$
Residual	SSE	n-k	$s^2 = SSE/(n-k)$	
Total	$SST = \Sigma y_{ij}^2 - SS(\mu)$	n-1		

have kept the sum of squares due to the mean in all the calculations. However in practice the sum of squares for the mean is removed at the start, as we have done in the ANOVA table.

Standard deviations of the t's can be found more easily by expressing t as a function of treatment means than by using regression formula. Alternatively, it can be expressed as a linear function of the individual observations as below:

$$t_i = \bar{y}_i - \bar{y}$$

$$= (1/r_i - 1/n)\times(\text{the sum of } r_i \text{ observations on treatment } i)$$

$$- (1/n)\times(\text{the sum of } n-r_i \text{ observations not on treatment } i)$$

$$\text{var } t_i = (1/r_i - 1/n)^2 \, r_i \, s^2 + (1/n)^2 \, (n - r_i) \, s^2$$

$$= (1/r_i - 1/n) \, s^2 \tag{5.4.5}$$

Usually it is the difference between treatment means which is of interest rather than individual means, and its standard deviation is rather simpler to calculate because means for different treatments come from different experimental units, and so are independent.

$$\text{var}(t_i - t_j) = \text{var}(\bar{y}_i - \bar{y}_j)$$

$$= \text{var } \bar{y}_i + \text{var } \bar{y}_j$$

$$= \sigma^2/r_i + \sigma^2/r_j \tag{5.4.6}$$

5.5 TESTING THE ASSUMPTIONS

All the means of assessing goodness of fit and peculiar observation of Chapters 2 and 3 apply to experimental design models, but again the special nature of design models means that we must look out for special points. First, from Chapter 2, Section 8.1, the variances and covariances of the residuals are given by

$$\text{var } e_{ii} = \left((r_j - 1)/r_j\right)\sigma^2 \quad \text{where unit } i \text{ receives treatment } j$$

$$\text{cov}(e_i, e_j) = 0 \quad \text{if units } i,j \text{ receive different treatments}$$

$$= (1/r_i)\sigma^2 \text{ if units } i,j \text{ receive the same treatment}$$

Second, in regression we wish to check whether the relationship between y and x is linear. This is not relevant when x is either 1 or 0. We do however wish to check whether the variance of y is the same in each treatment. After all, if a treatment can change the mean it can also change the variablity. This is best done by plotting a scatter plot of residuals against treatment number.

Finally, the plot of e against predicted y is just as relevant as in regression, to check whether the variance of y is a function of the expected value of y.

It is useful to indicate which treatment each point comes from in any residual plots. This is easily done by plotting each point using a different letter for each treatment instead of using "*" for every point.

Outliers will usually be difficult to explain away in an experiment because there will be more supervision of the experimental units than in regression based on descriptive data. Nevertheless they must still be identified and investigated. Also, because in an experiment the X matrix is designed, there should not be any high leverage points (Chapter 4, Section 2) nor should ridge regression techniques (Chapter 4, Section 7) be needed. Transformations (Chapter 4, Section 5) are every bit as important though, and should be considered if the variance of the observations does not seem to be constant.

Finally, experimental units are often likely to be correlated. For example they may be adjacent plots of ground or consecutive runs of a machine. A plot of residuals against order, in addition to the tests described in Chapter 2, Section 7 should check this. Randomization of treatments ensures that the parameter estimates will be independent of each other in spite of correlations between experimental units.

We conclude this chapter with an example of a computer analysis. The data is from Example 5.3.1, and the computer program used is GENSTAT. The analysis is shown in Figure 5.5.1, and a residual plot is shown in Figure 5.5.2. The output is almost self explanatory. The three STANDARD ERRORS OF DIFFERENCES OF MEANS are, respectively, for the difference between two means of two observations, the

***** ANALYSIS OF VARIANCE *****

SOURCE OF VARIATION	DF	SS	MS	VR
UNITS STRATUM				
TRT	2	0.70714	0.35357	4.562
RESIDUAL	4	0.31000	0.07750	
TOTAL	6	1.01714	0.16952	

***** TABLES OF MEANS *****

TRTMNT	A	B	C
	10.40	11.15	10.85
REP	3	2	2

***** STANDARD ERRORS OF DIFFERENCES OF MEANS *****

TABLE	TRTMNT	
REP	UNEQUAL	
SED	0.278	MIN REP
	0.254	MAX-MIN
	0.227X	MAX REP

(NO COMPARISONS IN CATEGORIES WHERE SED MARKED WITH AN X)

***** STRATUM STANDARD ERRORS AND COEFFICIENTS OF VARIATION *****

STRATUM	DF	SE	CV%
UNITS	4	0.278	2.6

FIGURE 5.5.1 GENSTAT analysis for simple example.

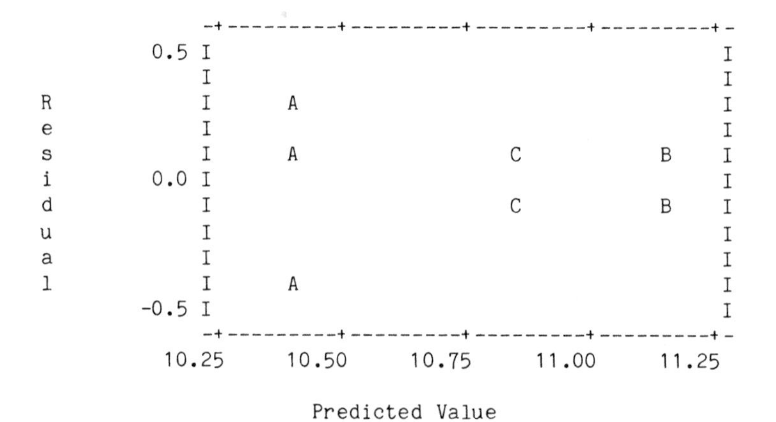

FIGURE 5.5.2 Plot of residuals for simple example.

difference between a mean of two observations and a mean of three, and the difference between two means of three observations. When there are several treatments all of different replication every pair might have a different standard error, but for most purposes the maximum, minimum and middle values given here are sufficient. The STRATUM STANDARD ERRORS give details of the estimate of σ from the ANOVA table. The CV% is the ratio of the SE to the overall mean, a useful measure of the precision of an experiment when percentage changes are important.

PROBLEMS

5.1. If the tea tasting experiment were arranged so that cups were tasted in pairs, how many pairs would be required for it to be possible for a 5% significance level to be achieved?

5.2. In the following cases identify the experimental units and explain what treatments are really being compared. How many replications are there in each case? Suggest ways in which the experiment might have been improved.

(i) Twenty patches of barley grass (a noxious weed) are identified in a paddock. To these four weed killers treatments are assigned randomly (three different weedicides and an untreated control), five patches per treatment. Four quadrats were randomly placed on each patch and an assessment was made of the barley grass cover within each.

(ii) Controlled climate rooms maintain levels of lighting, temperature and humidity to very close limits. Two such rooms were used in an experiment to study the growth of a tropical plant. One room was at high humidity, the other at low humidity. Other factors studied were the severity of pruning and the degree of watering. Each plant grew in a single pot and these pots were arranged as in the diagram below. The letters A, B, C, D denote four different severities of pruning, and the letters x, y, z the three different degrees of watering. Pots on the same watering treatments were kept together to minimise the amount of tubing required.

xA	xB	xD	xA
xB	xD	xC	xC
yA	yA	yD	yC
yD	yB	yC	yB
zB	zC	zA	zD
zC	zD	zA	zB

High humidity

yB	yC	yA	yB
yA	yD	yC	yD
xB	xD	xA	xC
xD	xC	xB	xA
zB	zA	zD	zC
zB	zC	zA	zD

Low humidity

(iii) A soil scientist randomly applied eight fertiliser treatments to 32 plots of ground. From each one he took four soil samples which were then bulked together to one pile of dirt. This pile of dirt was then churned up with chemicals to extract the elements being measured and two samples were taken from the resulting fluid, each being given an independent chemical analysis. The final data was then two numbers for each plot.

(iv) A class of fifty psychology students performed an experiment to assess the effect of violence in films. The class was split randomly into two groups of 25, and the first group watched a rather violent film after which each student completed a questionnaire. Then the second group watched the same film, but with the violent episodes cut, after which they also completed the questionnaire.

5.3. For the data in Example 5.3.1 calculate estimates under the restrictions (5.3.6) and (5.3.4). Calculate the residual sum of squares in each case to satisfy yourself that it really is the same regardless of the restriction.

5.4. For Example 5.3.1 show that the estimate of a treatment difference is the same regardless of what restrictions are used, but the estimate of the mean changes. Also show that the mean is not estimable, but that a treatment difference is.

5.5. A mob of sheep were randomly allocated to three dosing regiemes. Their weight gains (kg) for the three months of dosing were:

Trt A (not dosed):	-3	2	5	4	6	-1	-4	5	6	1	2		
Trt B (brand X) :	10	6	4	3	8	7	-2	1	8	6	7	4	-1
Trt C (brand A) :	0	8	9	-1	9	5	11	5	7	2	8	4	7

(i) Calculate the ANOVA table and test the hypothesis that dosing regieme had no effect on weight gains.

(ii) Calculate the standard deviation of the estimate of the differences between each pair of treatment means.

(iii) Comment on any suspicious features of the data.

5.6. The following GENSTAT output is from the experiment described in Appendix C 6. The difference between the two scores is analysed. Here the order factor has been ignored, leaving four treatments:

A: Focus method and computer taught.
B: Pairing method and computer taught.
C: Focus method and human taught.
D: Pairing method and human taught.

(i) Was the F-test significant at any reasonable significance level?

(ii) For children taught by computer what was the difference in word recognition performance between the two methods of teaching? What is the standard deviation of this estimate? Calculate a 90% confidence interval for the difference?

(iii) Comment on the scatter plot of residual against predicted value. Do the assumptions for ANOVA appear satisfied?

```
***** ANALYSIS OF VARIANCE *****

VARIATE: DIFFER

SOURCE OF VARIATION              DF          SS          MS        VR

*UNITS* STRATUM
  Method                          3      26.980       8.993     1.720
  RESIDUAL                       45     235.224       5.227
TOTAL                            48     262.204       5.463

GRAND MEAN                                1.53
TOTAL NUMBER OF OBSERVATIONS              49

***** TABLES OF MEANS *****

      Method        A       B       C       D

                 1.00    0.77    1.75    2.67
         REP       12      13      12      12

***** STANDARD ERRORS OF DIFFERENCES OF MEANS *****

TABLE              Method
---------------------------
REP               UNEQUAL
SED                 0.933  MIN REP
                    0.915  MAX-MIN
                    0.897X MAX REP

(NO COMPARISONS IN CATEGORIES WHERE SED MARKED WITH AN X)
```

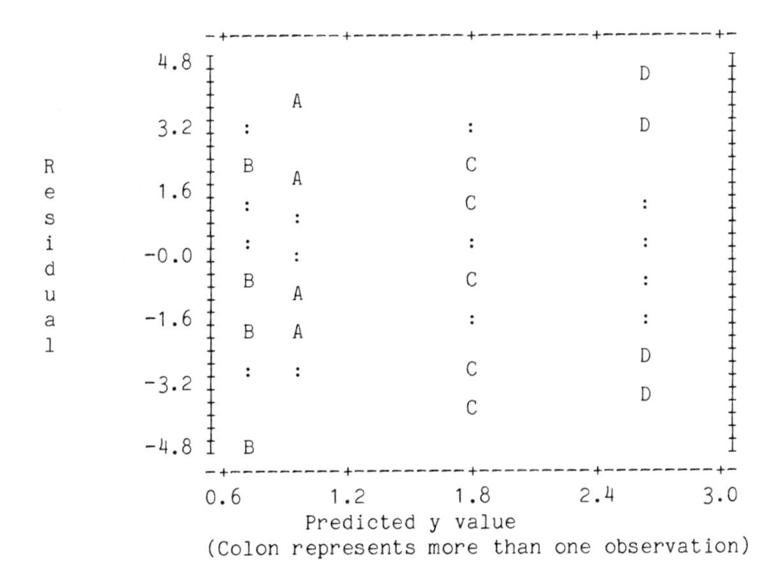

Predicted y value
(Colon represents more than one observation)

6
ASSESSING THE TREATMENT MEANS

6.1 INTRODUCTION

In the last chapter we described the linear model for a simple experiment, found estimates of the parameters, and described a test for the hypothesis that there is no overall treatment effect. In this chapter we cover the next step of examining more closely the pattern of differences among the treatment means. There are a number of approaches. One extreme is to test only the hypotheses framed before the experiment was carried out, but this approach wastes much of the information from the experiment. On the other hand, to carry out conventional hypothesis tests on every effect that looks interesting can be very misleading, for reasons which we now examine.

There are three main difficulties. First, two tests based on the same experiment are unlikely to be independent. Tests will usually involve the same estimate of variance, and if this estimate happens to be too small, every test will be too significant. Further, any comparisons involving the same means will be affected the same way by chance differences between estimate and parameter. As an example consider a case where there are three treatments assessing a new drug. Treatment "A" is the placebo, treatment "B" the drug administered in one big dose and treatment "C" the drug administered in two half doses. If chance variation happens to make the number of cures on the experimental units using the placebo (A) rather low, the differences between A and B, and A and C will both be overstated in the same way. Therefore two significant t-tests, one between A and B, the other between A and C, cannot be taken as independent corroboration of the effectiveness of the drug.

Second, picking out the largest effects from the many possible effects in an experiment clearly invalidates any simple hypothesis test applied to them. Picking two people from a large group and finding them both taller than 1.85m (73in) would be a rather surprising event if they were picked at random, but not at all surprising if they were the two tallest in the group.

Finally, any experiment will give rise to a great many possible inferences. It would be an unusual experiment which could not somewhere produce a "significant" comparison if one looked hard enough.

In this chapter we will describe the most common techniques and explain what they do. The difficulties outlined are too complex to be overcome by rules. The only sensible approach is to interpret results with a clear understanding of the dangers of the technique being used. Every technique is always available, but its appropriateness must be judged in each case.

6.2 SPECIFIC HYPOTHESES

Any experiment should be designed to answer specific questions. If these questions are stated clearly it will be possible to construct a single linear function of the treatment means which answers each question. It can be a difficult for the statistician to discover what these questions are, but this type of problem is beyond the scope of this book. We will present some examples.

<u>Example 6.2.1</u> Drug comparison example

In the drug comparison experiment mentioned in Section 1 one question might be, is the drug effective? Rather than doing two separate tests (A v B and A v C), a single test of A against the average of B and C gives an unambiguous answer which uses all the relevant data. That is use

$$\bar{y}_A - (\bar{y}_B + \bar{y}_C)/2 \quad \text{with variance} \quad [1/r_A + (1/r_B + 1/r_B)/4]\sigma^2$$

Having decided that the drug has an effect the next question may be, how much better is two half doses than one complete dose. This will be estimated by the difference between treatment means for B and C. For inferences remember that σ^2 is estimated by s^2, and this appears with its degrees of freedom in the ANOVA table.

Example 6.2.2 Fertilizer experiment

In an experiment to study the effect of potassium on the sodium uptake from the soil into grass the following treatments were compared:

A: No fertilizer
B: 56 kg sodium
C: 56 kg potassium
D: 56 kg sodium + 56 kg potassium
E: 112 kg sodium

Figure 6.2.1 gives a graphical representation of these treatments. The horizontal plane represents the levels of fertilizer, and the vertical axis represents the sodium content of the grass. The treatments were applied to plots of ground, the grass grew, was cut, and its sodium content measured. There are various questions. One might be, how much does sodium fertilizer increase sodium content? This can be measured by comparing the average of A and C with the average of B and D, which is the average upward slope of the surface ABDC moving from AC to BD. That is

$$(\bar{y}_B + \bar{y}_D)/2 - (\bar{y}_A + \bar{y}_C)/2$$

So long as the potassium has the same effect on its own as it does

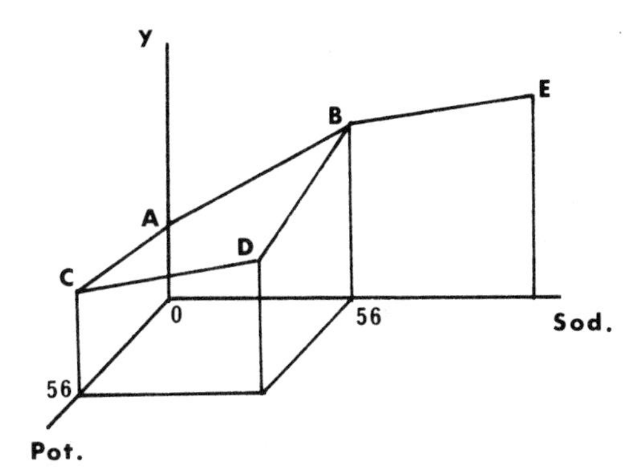

FIGURE 6.2.1 Treatments for fertiliser experiment.

when combined with sodium any difference between these pairs will be entirely due to the 56 kg of sodium. A comparison between the pairs A and B with C and D gives a measure of the potassium effect in a similar way. The effects being measured are called the main effects of sodium and potassium.

Example 6.2.3 Interaction

But potassium is a rather similar element to sodium, so perhaps it does not have the same effect on its own as it does when combined with sodium. Maybe the plant is not fussy whether it absorbs potassium or sodium, and if there is more potassium in the soil it absorbs less sodium. The increase in sodium content caused by the 56kg sodium applied to the soil will be less in those treatments where potassium is also present. To find out if this is happening, use

$$(\bar{y}_D - \bar{y}_C)/2 - (\bar{y}_B - \bar{y}_A)/2$$

Dividing by two has the practical purpose of ensuring that this function has the same variance as a main effect. The effect so measured is called the interaction between sodium and potassium. It is a measure of how twisted the surface ABDC is. If there is no interaction the surface will be a plane. Note that if the question were posed about the increase in sodium content caused by potassium in the presence or absence of sodium, the linear function would be exactly the same function. The interaction is symmetrical.

Example 6.2.4 Measure of nonlinearity

In any fertilizer response there tends to be a reduced response the higher the level. The difference in sodium content will be greater between the 56kg and none treatments than between the 112kg and 56kg treatments. Just how much greater is given by

$$(\bar{y}_B - \bar{y}_A) - (\bar{y}_E - \bar{y}_B) = -\bar{y}_A + 2\bar{y}_B - \bar{y}_E$$

Note that this is a measure of non linearity in response to sodium, or how far the points A, B and E depart from lying on a straight line.

Example 6.2.5 Steel coatings

Steels coated in four different ways were compared with two standard steels for making the drills of a plough. These were

1. Mild steel (the usual material).
2. Carbo-nitrided steel (a particularly hard material).
3. Steel top coated by process "A".
4. Steel bottom coated by process "A".
5. Steel top coated by process "B" .
6. Steel bottom coated by process"B" .

The six materials were used on 24 drills in a single plough, being positioned randomly. After some days continuous use a number of measurements were made to give an assessment of wear. A computer analysis using GENSTAT of the width across a drill is given in Table 6.2.1. Note that the F-ratio ("VR") is significant. This was expected because the mild steel and carbo-nitrided steel were known to

TABLE 6.2.1 Analysis of Plough Wear Experiment

***** ANALYSIS OF VARIANCE *****

SOURCE OF VARIATION	DF	SS	MS	VR
UNITS STRATUM				
COATING	5	393.675	78.735	12.887
RESIDUAL	18	109.970	6.109	
TOTAL	23	503.645	21.898	

***** TABLES OF MEANS *****

GRAND MEAN 79.07

COATING	1	2	3	4	5	6
	75.37	85.50	79.97	74.80	75.92	82.87

***** STANDARD ERRORS OF DIFFERENCES OF MEANS *****

TABLE	COATING
REP	4
SED	1.748

***** STRATUM STANDARD ERRORS AND COEFFICIENTS OF VARIATION *****

STRATUM	DF	SE	CV%
UNITS	18	2.472	3.1

be two extremes. The real question was how well coatings 3 to 6 performed. First of all, do they as a group wear better than mild steel? This is measured by

$$75.37 - (79.97 + 74.80 + 75.92 + 82.87)/4 = -3.02$$

whose estimated standard deviation is

$$\sqrt{((1 + (4/16))\times(6.109/4))} = 1.38$$

The t-statistic, with 18 degrees of freedom, is -3.02/1.38 = -2.19 which is significant at a 5% level.

Second, the two types of coating combined with top/bottom form a 2×2 pattern as in Example 6.2.2. Is top coating different from bottom coating? Use

$$(79.97 + 75.92)/2 - (74.80 + 82.87)/2 = -0.89$$

whose estimated standard deviation is

$$\sqrt{((2/4 + 2/4)\times(6.109/4))} = 1.24$$

The t-statistic, with 18 degrees of freedom, is -0.89/1.24 = -0.72 which is not at all significant.

The interaction (Example 6.2.3) assesses whether top coating or bottom coating is more important with one type of coating than the other. It is measured by

$$(79.97 - 74.80)/2 - (75.92 - 82.87)/2 = 6.06$$

Using the estimated standard deviation gives a t value of 6.06/1.24 = 4.89. This is a significant value, and a look at the data shows that top coating wore better than bottom coating with process "A", but bottom coating wore better with process "B". Because of this interaction the main effects give a misleading summary of the treatment effects.

6.2.1 Experimentwise Error Rate

The above are examples of inferences to answer specific questions. Each individual inference will be correct, but there are several inferences being made on each experiment. If all four suggested comparisons were made on the fertilizer experiment, the probability of making at least one type I error will be much higher than the significance level of an individual test. If the traditional 5% level is used, and there are no treatment effects at all, and the individual tests were independent, the number of significant results from the experiment would be a binomial random variable with $n=4$ and $p=.05$. The probability of no significant results will be $(1-0.05)^4$, so that the probability of at least one will be $1 - (1-0.05)^4 = 0.185$. If one really wanted to have the error rate per <u>experiment</u> equal to 0.05 each individual test would have to use a significance level, p, satisfying

$$1 - (1 - p)^4 = 0.05$$

$$\text{or} \qquad p = 0.013$$

Unfortunately, the underlying assumptions are false because, as we noted in Section 1, each inference is not independent. The correlation between test statistics will usually be positive because each depends on the same variance estimate, and so the probability of all four being nonsignificant will be greater than that calculated above and so the value of p given above will be too low. If the error rate per experiment is important, the above procedure at least provides a lower bound. Usually though it is sufficient to be suspicious of experiments producing many significant results, particularly if the variance estimate is based on rather few degrees of freedom and is smaller than is usually found in similar experiments. Experimenters should not necessarily be congratulated on obtaining many significant results.

In Section 1, another source of dependence was mentioned. This results from the same treatment means being used in different comparisons. If the questions being asked are themselves not independent, the inferences cannot be either. However, it is possible to design a treatment structure so that independent questions can be assessed independently. This will be the topic of the next section.

6.3 CONTRASTS

The linear functions of treatment means in Examples 6.2.1 to 6.2.5 were constructed because they were more relevant to the questions asked by the experimenter than the individual means. In constructing them the treatment parameters, the τ's, have been transformed to new parameters, which we will call γ's. The functions defining these γ's have one feature in common. If they were written in the form $\Sigma c_i \tau_i$, in each case $\Sigma c_i = 0$. Such functions are called <u>contrasts</u>. Formally then, if g is an estimate of a contrast γ, we have

$$\gamma = \Sigma c_i \tau_i = \mathbf{c}^T \boldsymbol{\tau} \tag{6.3.1}$$

$$g = \Sigma c_i t_i = \mathbf{c}^T \mathbf{t} \tag{6.3.2}$$

where $\Sigma c_i = 0$

Some general properties of estimates of these contrasts are

(i) g is an unbiased estimate of γ

$$E \sum c_i \bar{y}_i = \sum c_i(\mu+\tau_i) = \Sigma c_i + \Sigma c_i \tau_i = 0 + \gamma$$

It follows that a contrast is always estimable.

(ii) The variance of $g = \sum (c_i^2/r_i)\, \sigma^2$. (6.3.4)

(iii) Inferences about γ can be made using the t-statistic

$$g \,/\, \sqrt{(\Sigma(c_i^2/r_i)\, s^2)} \qquad (6.3.5)$$

where s^2 and its degrees of freedom come from the ANOVA table.

(iv) If two contrasts γ and γ' satisfy $\Sigma(c_i c_i'/r_i) = 0$, then the estimates of them, g and g', are independent random variables. This can be shown by expanding cov(g, g') and showing that it equals $\Sigma(c_i c_i'/r_i)\sigma^2$. If this is zero, g and g' are independent, since they are normally distributed random variables.

The independence of g and g' as random variables is quite different from the independence of γ and γ' as functions of the vector of treatment parameters, $\boldsymbol{\tau}$. In the drug comparison experiment (Example 6.2.1) two questions were asked, one about the overall response to the drug, and the other about the effect of the way it was administered. The two contrasts answer these questions independently, in the sense that any change in the overall response to the drug would change τ_B and τ_C equally, so that the contrast measuring the way it was administered will not be affected. Similarly in the fertilizer experiment, Example 6.2.2. An increase in the sodium effect will increase τ_B and τ_D equally, and this will not affect the size of the potassium or the interaction contrasts.

This indepence is a consequence of the functions defining the contrasts being orthogonal (see Appendix A 1.6). The coefficients satisfy $\Sigma c_i c_i' = 0$.

If there are k treatments, it is always possible to construct k-1 orthogonal contrasts. Not all of these may have any physical meaning, but it will simplify the discussion if all are included. Geometrically, the vector space representing treatments is being split into k-1 orthogonal components. The parameters are being transformed from $\boldsymbol{\tau}$ to $\boldsymbol{\gamma}$ using the orthogonal matrix C whose rows are the coefficients defining each contrast, so that $\boldsymbol{\gamma} = C\boldsymbol{\tau}$. Also, following (5.3.6), the τ's add to zero. The transformation can be written

$$\begin{vmatrix} 0 \\ \cdots \\ \boldsymbol{\gamma} \end{vmatrix} = \begin{vmatrix} \mathbf{1}^T \\ \cdots \\ C \end{vmatrix} \boldsymbol{\tau} \qquad (6.3.7)$$

If the rows of the matrix are normalized, so that $\Sigma c_i{}^2 = 1$, C becomes orthonormal. That is, its inverse is its transpose. From (6.3.7) then,

$$(\mathbf{1} , C^T) \begin{vmatrix} 0 \\ \cdots \\ \boldsymbol{\gamma} \end{vmatrix} = C^T \boldsymbol{\gamma} = \boldsymbol{\tau}$$

This value for $\boldsymbol{\tau}$ can now be substituted in the design model, which becomes

$$\mathbf{y} = \mathbf{1} + X C^T \boldsymbol{\gamma} + \boldsymbol{\varepsilon} \qquad (6.3.8)$$

Thus the model is now expressed in terms of parameters directly related to the interests of the experimenter. Each column of XC^T comprises the coefficients of a contrast, the i-th coefficient appearing r_i times. If each treatment is equally replicated, the columns will be orthogonal because the contrasts are. We are then in the situation of Chapter 3, Section 2. The results of that section applied here give

(i) Each contrast is estimated independently and the drop in residual sums of squares caused by adding any contrast to the model is independent of any other contrast in the model.

(ii) The model sum of squares for (6.3.8), which is of course the treatment sum of squares, equals

$$\mathbf{g}^T C X^T X C^T \mathbf{g} = r \, \Sigma g_i^2 \qquad (6.3.9)$$

(iii) The sum of squares for the i-th contrast, $SS(\gamma_i)$ is $rg_i{}^2$.

(iv) The hypothesis H: $\gamma_i = 0$ can be tested using

$$F(1,n-k) = SS(\gamma_i)/s^2 = rg_i^2/s^2 \qquad (6.3.10)$$

Note that this is the square of the t-statistic (6.3.5).

(v) In practice it is a nuisance normalizing the contrasts. If g' is any contrast, the corresponding normalized contrast g and its sum of squares are given by

$$g = g'/(\Sigma c_i^2)$$

$$SS(\gamma') = r\, g'^2/\Sigma c_i^2$$

(vii) An experimenter may consider that only a few of the contrasts should be needed to explain all the treatment effects. The difference between the sum of squares for these contrasts and the treatment sum of squares corresponds to lack of fit, as described in Chapter 2, Section 10.

6.3.1 Regression Contrasts

In Example 6.2.4 reference was made to a measure of the nonlinearity of response to sodium. Experiments very often do compare treatments which differ quantitively - different rates of fertilizer, different concentrations of reagents, different speeds of processing, different times of instruction - and it is then natural to fit a regression model to the data. We will show that a regression coefficient corresponds to a contrast. Consider the following example.

Example 6.3.1 Rate of weedicide

The experiment described in Appendix C 7 included amongst its eleven treatments a control (no treatment) and three rates (0.5, 1.0 and 1.5 litres) of the weedicide. One of the aims of the experiment was to see if the weedicide affected pea yields. The weedicide might improve yields by killing the weeds and lessening the competition, or it might worsen yields by killing the peas. Indeed there may be an optimum rate sufficient to kill the weeds without affecting the peas. These points can best be assessed by regressing yield on rate. Even if a strictly linear relationship is not expected, an hypothesis of no linear regression specifically denies any monotonic change of yield with rate and is therefore a more powerful test than the overall F-test. A large quadratic regression coefficient would suggest that an optimum rate had been exceeded. The first step should be to draw a graph of yield against rate. It is shown in Figure 6.3.1.

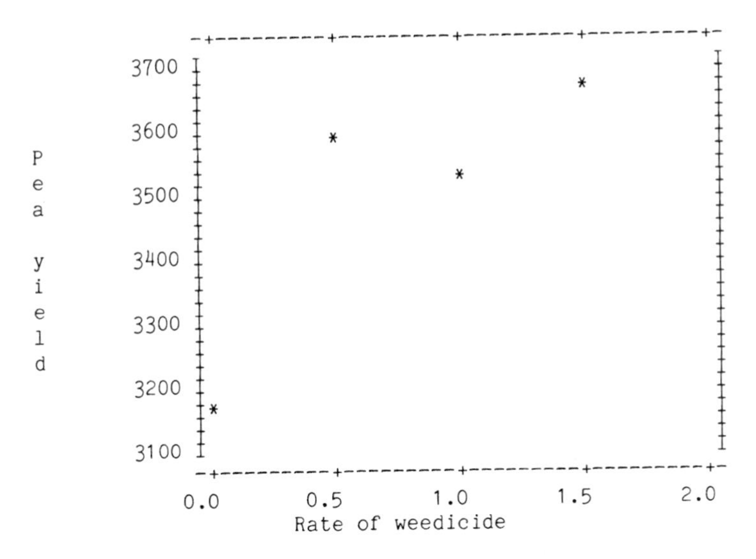

FIGURE 6.3.1 Graph of average pea yield against weedicide rate.

The usual design model

$$y_{ij} = \mu + \tau_i + \varepsilon_{ij} \qquad 1 \leq i \leq 4$$

is replaced with

$$y_{ji} = \beta_0 + \beta_1 x_{ji} + \beta_2 x_{ji}^2 + \beta_3 x_{ji}^3 + \varepsilon_{ji}$$

where x_{ji} = i-th rate, the same value for each j. Note that these two models give X matrices of the same rank. From Chapter 3, Section 4 we know that the x's can be replaced by z's which are orthogonal. First the linear term can be made orthogonal to the constant by subtracting its mean. To make its coefficients integers it can be multiplied by four. The calculations are shown below

Rate (x_i) :	0.0	0.5	1.0	1.5
($x_i - \bar{x}$) :	-0.75	-0.25	0.25	0.75
$z_{1i} = 4(x_i - \bar{x})$:	-3	-1	1	3

The estimate of z_1's coefficient is

$$b_1 = [\sum (z_{1i} y_{i.})]/[r \sum z_{1i}^2] \qquad \text{from (2.7.2)}$$
$$= \sum (z_{1i} \bar{y}_{i.})/20$$

This is a contrast with $c_i = z_i/20$. Indeed, looking back now, we can see that any contrast is really the regression coefficient for a regression of the treatment means against the c_i's. Also x^2 and x^3 can be transformed to z_2 and z_3 by following the procedures of Chapter 3, Section 4. The results of the analysis are given in Table 6.3.1. The linear effect is significant, but the quadratic and cubic effects are not. Note that the overall treatment sum of squares is not significant, so that the linear regression contrast has found an effect which might otherwise have been missed.

The above procedure can be followed to fit any range of x values. If the x's are evenly spaced the z's can be found from tables of orthogonal polynomials and these simplify hand calculations considerably. However, most computer programs enable regression effects to be fitted for any x's, evenly spaced or not, so orthogonal polynomials are not as important as they once were. One common special case arises when y is believed to be proportional to the log of x. If the rates of x are chosen to be in geometric progression their logs will be in arithmetic progression. The model

TABLE 6.3.1 Calculation of Regression Contrasts

Rate:	0.0	0.5	1.0	1.5	b_i	$SS(b_i)$	F
Yield means:	3155	3573	3515	3657			
Linear contrast	-3	-1	1	3	72.4	$4 \times 72.4^2 \times 20$ $= 419341$	6.1
Quadratic contrast	1	-1	-1	1	-69.0	76176	1.1
Cubic contrast	1	-3	3	-1	-33.8	91395	1.3
Overall treatment						586912	2.8

$y = \beta_0+\beta_1(\log x)$ can then be fitted by calculating the z's as in the previous example. (e.g. if $x = 1,2,4,8$ then $z = -3,-1,1,3$)

Although we have explained how to find cubic contrasts, it is most unlikely that any useful interpretation can be found for them or indeed even for quadratic contrasts. In this example one may wonder whether more than linear terms need be include in the model. Here a lack of fit test is appropriate. The sum of squares for lack of fit is $SS(\tau)-SS(\beta_1)$ which has 2 degrees of freedom. The F-test for lack of fit is given by:

$$\{[SS(\tau) - SS(\beta_1)]/2\}/s^2 \sim F_{2,30}$$

In our example this is $\{[586912-419341]/2\}/68643 = 1.22$, which is not significant.

It is well worth remembering that none of the statistics we have calculated display the relationship as well as Figure 6.3.1.

6.4 FACTORIAL ANALYSIS

Main effects and interactions have been introduced as contrasts. However, there is no reason why a main effect should not have several levels. The following is an example of such an experiment.

<u>Example 6.4.1</u> Milk fat sampling

Anhydrous milk fat is sampled for quality tests, but the samples drawn can themselves deteriorate. Two factors which could affect the sample quality are the container used and the way the containers are filled. An experiment to study these two factors used four different containers: metal tins, glass jars, clear plastic jars, opaque plastic jars. These were filled using two different methods: from the bottom and from the top. Four samples were taken for each of the eight combinations of the two factors giving what is called a <u>factorial</u> experiment. The samples were then sealed and after three days their peroxide level (a measure of oxidation) was measured. The results given in Table 6.4.1 are increases in peroxide values after a

TABLE 6.4.1 Data for Method of Filling Experiment

	Metal tins	Glass jars	Plastic jars		Means
			Clear	Opaque	
From bottom	0.02	0.04	0.06	0.04	0.040
From top	0.02	0.08	0.09	0.05	0.060
Means	0.020	0.060	0.075	0.045	0.05

three day period following sampling. Although this experiment could be treated as a simple, eight treatment experiment, questions about the results are going to fall into three categories: questions about the filling method, questions about the container, and questions about the interaction. A natural way, then, to express the treatment effect is to define τ_{lm} as the effect of the l-th container with the m-th filling method, and put

$$\tau_{lm} = \alpha_l + \beta_m + (\alpha\beta)_{lm} \qquad 1 \leq l \leq u,\ 1 \leq m \leq v \tag{6.4.1}$$

Thus the effect of any individual treatment comprises a container effect plus a filling method effect plus an effect unique to the particular combination. The first two are main effects and the last is an interaction. Forming estimates of these effects involve the same problems discussed in Chapter 5, Section 3. We will follow (5.3.6) and impose restrictions

$$\Sigma\alpha_l = \Sigma\beta_m = \sum_l(\alpha\beta)_{lm} = \sum_m(\alpha\beta)_{lm} = 0 \tag{6.4.2}$$

This is equivalent to defining the effect of a level of a treatment as the average change it causes from the overall mean, so that

$$\alpha_l = \bar{\tau}_{l\cdot} \tag{6.4.3}$$

$$\beta_m = \bar{\tau}_{\cdot m}$$

$$(\alpha\beta)_{lm} = \tau_{lm} + \tau_{l\cdot} - \tau_{\cdot m} \tag{6.4.4}$$

Estimates are obtained by substituting the appropriate estimates of the τ's. Those obtained for Example 6.4.1 are shown in Table

TABLE 6.4.2 Method of Filling Experiment : Estimates of Effects

	Metal tins $(\alpha\beta)_{1j}$	Glass jars $(\alpha\beta)_{2j}$	Plastic jars clear $(\alpha\beta)_{3j}$	Plastic jars opaque $(\alpha\beta)_{4j}$	Filling method β_j
From bottom $(\alpha\beta)_{i1}$	0.010	-0.010	-0.005	0.005	-0.01
From top $(\alpha\beta)_{i2}$	-0.010	0.010	0.005	-0.005	0.01
Container α_i	-0.030	0.010	0.025	-0.005	0.05 (μ)

6.4.2. This table shows that metal tins had the lowest peroxide level (0.03 below average) and clear plastic the highest (0.03 above average). The estimate of the mean for filling metal tins from the bottom is calculated by adding the interaction term in the body of the table (0.01), the main effects from the margins (-0.03 and -0.01) and the overall mean (0.05) giving 0.02, the treatment mean again. The value of this procedure lies not in its ability to recover treatment means, but in the way it can give estimates of treatment parameters when one or more of the effects is considered to be zero. Such effects would just be omitted from the addition. Each step further into the table (from overall mean, to margins, to the body) represents a refinement to the model and it does not make much sense to include a term unless all the terms further outwards from it are also included. For example, the overall mean is further outward than any other term, and it is the first term to include in the model. Also if a model includes an interaction it should also include the main effects. Terms "further outward from" are said to be *marginal to*.

If this type of estimate were used in Examples 6.2.2 and 6.2.3, the magnitude of the effects would be half the magnitude of the contrasts, because the contrasts were based on the difference between the level means instead of the difference between a level mean and the overall mean.

If every treatment has equal replication, every contrast among the α's is orthogonal to every contrast among the β's. Questions about different main effects can therefore be answered independently.

If $\gamma_a = \Sigma c_l \alpha_l$ is a contrast among the α's

$$\gamma_a = \Sigma c_l \alpha_l$$

$$= \Sigma c_l \bar{\tau}_{l.}$$

$$= \sum_{ml}\Sigma c_l \tau_{lm}/2$$

and similarly if $\gamma_b = \Sigma c'_m \beta_m$ is a constrast among the β's

$$\gamma_b = \sum_{lm}\Sigma c'_m \tau_{lm}/4$$

To test for orthogonality, we evaluate

$$\sum_l \sum_m (c'_m/4)(c_l/2) = (\Sigma c'_m)(\Sigma c_l)/8 = 0$$

It can also be shown that contrasts among the $(\alpha\beta)$'s are orthogonal to both the above. The k-1 dimensional space for the full model has been split into three mutually orthogonal subspaces, one for each main effect, and one for the interaction. This means that the three groups of parameters, α's, β's and $\alpha\beta$'s, can be estimated and tested independently.

$SS(\alpha)$ is obtained from the model

$$y_{lj} = \mu + \alpha_l + \varepsilon_{lj} \qquad \text{(8 y's for each l)}$$

which is really just the standard model with α instead of τ and 8 observations at each level. Therefore

$$SS(\alpha) = 8\,(\bar{y}_{l.} - \bar{y}_{..})^2$$

$$= 8\,\bar{y}^2_{l.} - 32\,\bar{y}^2_{..} \qquad (6.4.5)$$

$$SS(\beta) = 16\,\bar{y}^2_{.m} - 32\,\bar{y}^2_{..}$$

$$SS(\alpha\beta) = SS(\tau) - SS(\alpha) - SS(\beta) \qquad (6.4.6)$$

It is useful to note that the sum of squares of a factor is always of the form

$$\sum \text{(Replication of each level)}\times\text{(level mean)}^2 - n\times\text{(mean)}^2 \quad (6.4.7)$$

where the summation is over all levels of the factor. Using this, the sums of squares for experiments with three or more factors can be calculated. For example if there are three factors α, β and γ, the treatment sum of squares can be partitioned into

$$SS(\tau) = SS(\alpha) + SS(\beta) + SS(\gamma) + SS(\alpha\beta) + SS(\alpha\gamma) + SS(\beta\gamma) + SS(\alpha\beta\gamma) \quad (6.4.8)$$

The two factor interaction sums of squares for $(\alpha\beta)$ are calculated by (6.4.6) where $SS(\tau)$ is replaced by the sum of squares from the means for $(\alpha\beta)$ taken over reps and the levels of γ. This quantity we will label $SS(\alpha{*}\beta)$ and (6.4.6) can be restated

$$SS(\alpha\beta) = SS(\alpha{*}\beta) - SS(\alpha) - SS(\beta)$$

Such calculations are straightforward in principle but tedious in practice, so are best left to a computer.

Example 6.4.1 Continued

The data from Table 6.4.1 are now analysed. Note that it is most important to keep high accuracy when calculating differences between squares. Rule (6.4.7) is used to form all the sums of squares.

Total sum of squares
$= 0.0250$ (from individual results not quoted)

Treatment sum of squares
$= 4\times(0.02^2 + 0.04^2 + \cdots + 0.05^2) - 32\times0.05^2 = 0.0184$

Container sum of squares
$= 8\times(0.020^2 + \cdots + 0.040^2) - 32\times0.05^2 = 0.0132$

Filling sum of squares
$= 16\times(0.04^2 + 0.06^2) - 32\times0.05^2 = 0.0032$

Interaction sum of squares
$= 0.0184 - 0.0032 - 0.0132 = 0.0020$

All these quantities can be used to form an ANOVA, as in Table 6.4.3.

TABLE 6.4.3 ANOVA for Milk Storage Experiment

Source	S. Sqs	D.F.	M.S.Sqs.	F
Treatments	0.0184	7	0.00262	9.5 **
Container	0.0132	3	0.00440	16.0 **
Filling	0.0032	1	0.00320	11.6 **
Interaction	0.0020	3	0.00067	2.4
Residual	0.0066	24	0.00028	
Total	0.0250	31		

The conclusion is that both containers and filling methods affect the peroxide level. The main effects shown in Table 6.4.2 are significant, but the lack of interaction indicates that the increase caused by the filling method is uniform over all the containers and the effects in the body of Table 6.4.2 could well be chance variation.

6.4.1 Unequal Replication

Experimenters in the social sciences can rarely achieve equal replication of all treatments. What do they lose? Primarily, unambiguous interpretation. If replication is not equal, orthogonal contrasts are not independent, and model (6.3.8) does not have orthogonal columns. In a factorial experiment the columns of X corresponding to different factors are not orthogonal. The estimates of contrasts within one factor therefore depend on whether the other factor is in the model. Consequently we can no longer talk about _the_ sum of squares for a factor. The procedure involves fitting a series of models using general linear regression methods, and belongs better in the context of the final chapter.

6.5 UNPREDICTED EFFECTS

Every experiment will show effects which appear interesting, but which were not thought of beforehand. Some effects will be large by chance, and even if there are no true treatment effects the largest of the many possible effects might well reach a nominal 5% significance level. In the introduction to this chapter we considered the problem of picking two people from a large group and finding them both taller than 1.85m. The techniques in earlier sections of this chapter apply when the choice is made randomly, or at least by a process independent of heights. Our situation now is much more like picking the two tallest in the room and then wanting to know whether the group is especially tall. Therefore the methods of inference explained earlier are no longer appropriate.

Nevertheless, some measure of the size of effect which chance variability might cause is useful, and a principle to invoke is to consider the distribution (under the null hypothesis) of the maximum of all the effects of the same type as the one of interest.

Applying this principle to differences between treatment means leads to Tukey's Honest Significant Difference, (HSD). The effects of interest are the differences between means, and the maximum effect of this type is the range. If a particular pair of means differ by an amount d we assess whether d is extreme by evaluating

$$\text{pr}(\text{range} > |d|) \tag{6.5.1}$$

If $Q(f,k)$ is the range of an independent sample of size k from a normal population whose standard deviation is estimated by s with f degrees of freedom, tables giving the values of $Q(f,k;\ \alpha)$ such that

$$\text{pr}(\text{range}/s > Q) = \alpha \tag{6.5.2}$$

for selected values of α, can be found in Pearson and Hartley, 1966. For a set of k treatment means, each replicated r times, the standard deviation is estimated by $s/\sqrt{r}$ and $f = n-k$. We can therefore say quite exactly that the probability is 0.05 that, if the null hypothesis is true,

$$|\bar{y}_i - \bar{y}_j| > Q(n-k,\ k;\ 0.05)\ s\ /\ \sqrt{r} \tag{6.5.3}$$

for all pairs i and j, so that the statement applies to any particular pair however it was chosen. Compare this with the standard t test for a single specified contrast. The test statistic is t with n-k degrees of freedom:

$$|\bar{y}_i - \bar{y}_j| > t_{n-k}\, s \sqrt{(2/r)}$$

The HSD procedure gives an exact hypothesis test for the range which may be useful in some specialized situations. Modifications have been suggested to provide a multiple range test. A multiple range test purports to divide the treatments into groups within which there are no significant differences. Their use can be criticized on three grounds. First, experimenters rarely want to divide treatments into groups. They really want to estimate differences, not test them. Second, the tests are devised by making assumptions about the balance between errors per experiment and errors per comparison which make a nominal 5% significance test mean something very different from what most experimenters understand. Finally tests using contrasts are much more powerful if answers are required to specific questions.

The use of contrasts provide a second application of our principle, and leads to Scheffe's S statistic. The range was the maximum value for any treatment difference. The largest contrast which can be constructed is one which corresponds to a vector in the direction of the vector $X\mathbf{t}$. All the sums of squares for treatments will be concentrated in this contrast, so that the largest possible contrast sums of squares is the treatment sums of squares, $SS(\tau)$. This contrast can be constructed by taking $c_i = \bar{y}_i - \bar{y}_{..}$

From the ANOVA table the distribution of $[SS(\tau)/(k-1)]/s^2$ is $F(k-1,n-k)$. To discover whether a large contrast sum of squares, S_0, is significant we calculate

$$pr\{SS(\tau) > S_0\} = pr\{[F(k-1,n-k)\, s^2\, (k-1)] > S_0\} \qquad (6.5.5)$$

If F is the 95 percentile of $F(k-1, n-k)$ we have that

$$pr\{SS(\tau) > F s^2 (k-1)\} = .05 \quad (6.5.6)$$

We can say with assurance that the probability is .05 that, if the null nypothesis is true,

$$pr\{SS(\gamma) > F s^2 (k-1)\} = .05 \quad (6.5.7)$$

for <u>all</u> contrasts γ. Compare this with the standard F test for a single, specified contrast in which F is the 95th percentile of F(1, n-k), giving

$$pr\{SS(\gamma) > F s^2\} = .05$$

An example in the next section demonstrates these results.

6.5.1 A Graphical Method

Although the principle upon which the methods of the previous section is based is important, the techniques themselves are really the wrong tools. What is required is a method for looking at the results of an experiment and seeing where the differences lie.

It is a standard result of distribution theory that if the values of a random sample from any distribution are put in order from smallest to largest, their expected values divide the density curve into approximately equal areas. Therefore if y is the expected value of the i-th largest observation in a random sample of size n from a distribution with distribution function F, F(y) is approximately equal to i/(n+1). However F(y) = (2i-1)/2n gives a slightly better approximation.

If there is no difference between treatments in an experiment, randomization with the central limit theorem ensures that the means will be approximately equivalent to a random sample from a normal population. An estimate of the expected value of the i-th largest mean is therefore given by

$$x_i = \bar{y} + s z_i \quad (6.5.8)$$

where z_i satisfies $F(z_i) = (2i-1)/2n$ and F is the standard normal distribution function. A plot of the ordered treatment means against values of z gives the required picture. Groups of treatments within which there are no more than sampling differences will lie approximately parallel to the line (6.5.8). Where there are real differences the slope will be greater than this line. The technique is demonstrated in the following example.

Example 6.5.1

We will calculate some of the above statistics for the plough experiment, Example 6.2.5. First, Tukey's HSD uses Q(18,6;0.05) which is 4.49. The honest significant difference is

$$4.49 \times \sqrt{(6.109/4)} = 5.55$$

Compare this with the corresponding value for a t-test using 18 degrees of freedom:

$$2.101 \times \sqrt{(6.109(1/4 + 1/4))} = 3.672$$

As a difference is just a special case of a contrast, a Scheffe's test could be appropriate also. For a difference

$$SS = r \times (\text{difference})^2 / 2$$

and (6.5.6) can be rearranged to give

$$pr\{\text{difference}^2 > 2 [F s^2 (k-1)/r]\} = .05$$

A minimum significant difference corresponding to the above is given by

$$\sqrt{[2F s^2 (k-1)/r]} = \sqrt{(2 \times 2.77 \times 6.109 \times 5/4)} = 6.504$$

The t-test is appropriate if the difference was selected for reasons independent of its size, the HSD is appropriate if it was picked out as a large difference, and the Sheffe's test is appropriate in the unlikely event that it was picked out as a large constrast.

If treatments are ordered from smallest to largest it is possible to use underlines to indicate those groups of means which differ by less than these amounts, as in Table 6.5.2. You should by now appreciate that not one of these underlines really represents a precise inference in any sense. They are really only attempts to display patterns among the means. But even this is not done as well as the graphical method, which we demonstrate in Figure 6.5.1. The dotted line is the line (6.5.8), $y = 79.1 + 1.23x$. The graph shows that points 4, 1 and 5 lie in a line of slope parallel to the dotted line, so that the differences between these means are what would be expected from random variation. The big jump is to point 3, with rather smaller differences between 3 and 6 and 2. The picture displays the pattern very much better than the underlining.

6.5.2 Ranking and Selection Methods

This quite recent development is far too large for us to more than mention here. Instead of starting with the null hypothesis that there are no treatment effects, ranking and selection methods start from the assumption that there are treatment differences and test hypotheses about their order. Questions such as, which is the best treatment, are therefore answered directly. For a description of the

TABLE 6.5.2 Results of Multiple Range Tests

Treat. No:	4	1	5	3	6	2
Means	74.80	75.37	75.92	79.97	82.87	85.50
t-test	———	———	———	———	———	———
H S D	———	———	———	———		
				———	———	———
Scheffe	———	———	———	———		
				———	———	———
(2i-1)/12	1/12	3/12	5/12	7/12	9/12	11/12
$z_i = F^{-1}(2i-1)/12$	-1.38	-.67	-.21	+.21	+.67	+1.38

(t-test underlines: 4–1–5; 3–6; 2. H S D and Scheffe underlines: 4 through 3; 3 through 2.)

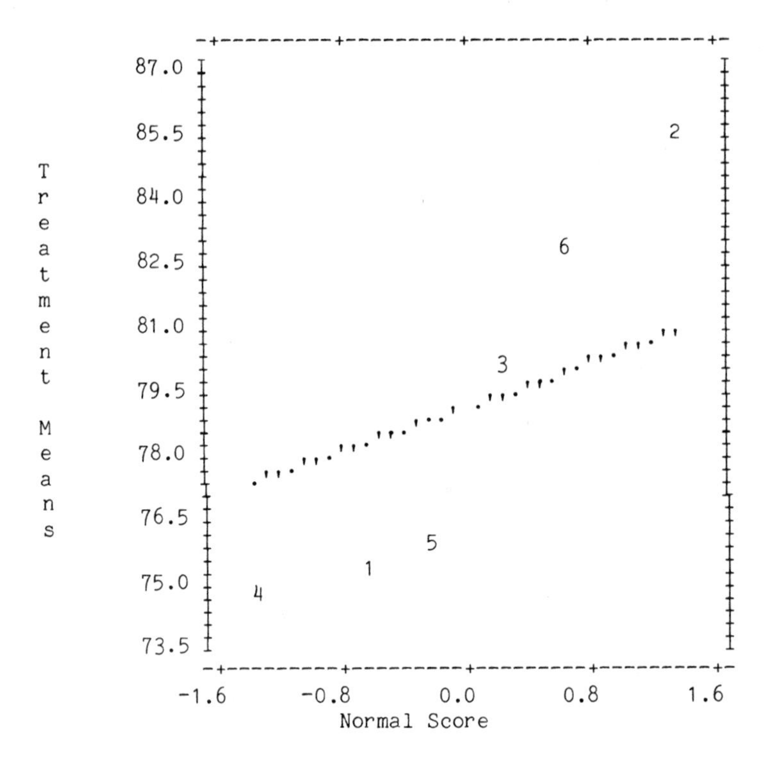

FIGURE 6.5.1 Ordered treatment means plotted against normal score.

method see Gibbons, Olkin and Sobel (1977). The calculations are complicated, but an interactive computer program, RANKSEL, is available to handle some of them (Edwards, 1984).

6.6 CONCLUSION

If any general principle is to be extracted from the preceeding detail perhaps the most important one is that the questions an experiment is expected to answer must be clearly thought out when the experiment is being designed. It is often possible to choose treatments in such a way that each question corresponds to a separate orthogonal contrast. After the experiment these can be tested and estimated, and further effects sought. The best procedure is to start with an F-test for overall treatments, then consider tests for main effects and interactions and finish with single contrasts. The harder one has to look, the less notice one should take of what is found. Graphs can provide a much better idea of the effects than any number of hypothesis tests.

Finally, one should never forget that for most purposes the basic question is, how much? That is, not an hypothesis test, but an estimation.

PROBLEMS

6.1. Once upon a time every second agricultural field experiment compared the effect of the elements N,P and K. One such experiment had these elements in the four combinations: None, N + P, P + K, K + N. Corn was grown on 24 plots, and yields (in old fashioned bushells/acre) were as below:

None	99	40	61	72	76	84
N + P	96	84	82	104	99	105
P + K	63	57	81	59	64	72
K + N	79	92	91	87	78	71

(i) Construct three orthogonal contrasts, one for measuring the effect of each element, and thus analyse the data. What assumption must be made for these contrasts to be measures purely of the effect of each element?

(ii) Calculate three confidence intervals, one for each contrast, such that one can be 90% confident that all three include the parameter they estimate.

6.2. For Question 5, Chapter 5, construct two contrasts, one measuring the effect of dosing, the other measuring the difference between the two brands. Show that they are orthogonal, but that their sums of squares do not add to the treatment sum of squares. Explain why not.

6.3. When whole milk is left standing in a vat the cream rises to the top, making sampling difficult. A rotating paddle can be used to mix the milk, but too much mixing can damage it. The following data is from an experiment to study how paddle size and speed of rotation affect the mixing time. There were two independent runs at three levels of both speed and paddle size.

Rotational speed (rpm)	Diameter (mm)	Mixing time (sec)	
75	90	1490	1100
75	116	780	620
75	150	825	690
110	90	575	480
110	116	385	385
110	150	315	400
160	90	195	230
160	116	117	190
160	150	225	145

(i) By plotting differences between observations on the same treatment against their mean, show that a log transformation is appropriate.

(ii) Calculate the ANOVA for these data using the log of the mixing time. Test both main effects and interaction for significance.

(iii) Show that because the three diameters are in geometric progression, a linear regression contrast using the x values -1,0,1 measures the linear relationship of log(diameter) against log(mixing time). Calculate this contrast. Is there evidence that it is not an adequate explanation of the diameter effect?

6.4. Carry out the following tests for the weedicide experiment, Appendix C 7:

(i) Plot the probability graph of the treatment means and interpret the result.

(ii) Test the differences between treatments A, B, C and D using HSD's. Should one treat these as a group of 4 means or a group of 11? Compare the result of these tests with the regression contrast test in Section 7.1.

(iii) Treatments A,C,J,I form a 2×2 pattern. Test the hypothesis that there is no interaction between "X" and Trietazine.

6.5. In Example 6.2.5 the interaction was found to be significant. Would it have been significant with Scheffe's test?

6.6 In the Example 6.3.1 coefficients for the quadratic and cubic contrasts were quoted. Confirm that they indeed are the correct values using the procedure of Chapter 3, Section 4.

7
BLOCKING

7.1 INTRODUCTION

In the last chapter we concentrated on the treatments of the experiment, and looked at ways of introducing structure into the treatment effects to answer specific questions. There are two other aspects of experimental design, and these will be introduced in this chapter. Often an experimenter can group experimental units in such a way that the units within groups are more likely to be uniform than units in different groups. Treatments can be allotted to take this grouping into account so that treatment comparisons are made between experimental units which are as uniform as possible. A typical experimental plan would be as in Figure 7.1.1. Here six treatments (perhaps six different crop varieties) are being compared on 24 experimental units which have been grouped into four blocks of six contiguous plots. This should seem sensible, because six contiguous plots are more likely to be uniform than the whole 24 plot area. Then treatments have been allocated to plots by making a block contain one complete set of treatments, or one replicate, randomly arranged within the block.

	Block 1						Block 2					
Plot No.	1	2	3	4	5	6	7	8	9	10	11	12
Treatment	A	B	E	F	D	C	D	E	C	A	B	F
	Block 3						Block 4					
Plot No.	13	14	15	16	17	18	19	20	21	22	23	24
Treatment	B	A	F	C	D	E	A	B	F	C	E	D

FIGURE 7.1.1. A simple experimental layout

7.2 STRUCTURE OF EXPERIMENTAL UNITS

The only structure we will be concerned with in this chapter is the one just described, where the experimental units are arranged in equal sized groups. If treatments are ignored the model for this structure is

$$y_{ij} = \mu + \xi_j + \varepsilon_{ij} \tag{7.2.1}$$

where y_{ij} is the measurement on the i-th unit in the j-th group. ξ_j is the effect of the j-th group, and ε_{ij} is the effect of the i-th unit within the j-th group. Both ξ and ε are random variables. The indices i and j are only labels and these labels can be arbitarily reassigned, j over the groups and i within groups, without changing the structure.

An earlier example of this structure arose in Problem 5.2(i). Here the groups were patches of barley grass, and within each patch measurements were made on four randomly placed quadrats. In that case it was the groups which were the experimental units, giving a very different sort of experiment from the one in Figure 7.1.1. However the difference lay not in the structure of the experimental units, but in the way the treatments were allotted to those experimental units. (7.2.1) really combines two linear models, one for the groups (indexed by j) and one for the individuals (indexed by i). In the computer output for block designs there will be reference to two <u>stratum</u>, one for each of these models. In this chapter we will be mainly concerned with experiments where treatments are randomly allotted within groups. The grouping is an attempt to eliminate any natural variability in the experimental units, and the possibility of the groups being experimental units will usually be ignored. The linear model will therefore be

$$y_{ij} = \mu + \beta_j + \varepsilon_{ij} \tag{7.2.2}$$

where the random ξ has been replaced by an ordinary parameter β. The sum of squares for the block stratum can be regarded as the sum of squares for a factor "blocks".

7.2.1 Randomized Complete Blocks

An experiment where each block comprises a complete set of treatments is called a randomized complete block. Figure 7.1.1 is such an experiment, and because it is easy to visualize, it is used in the following example.

Example 7.2.1 Artificial data

Artificial data is used in this example to demonstrate the effect of blocking. It has been generated by taking a constant (600), adding a sine function of the plot number to simulate a fertility trend, and adding some random variation. Then the treatments have been allocated to the plots in two ways, once to give a blocked experiment (as in Figure 7.1.1) and once to give an unblocked experiment. As we are interested only in the effect of grouping the experimental units, no treatment effects are added. The yield of each plot is therefore the same regardless of which treatment it receives. Any differences between treatment means are a result of the way treatments have been allocated to experimental units. Table 7.2.1 shows the allocation of treatments, and the artificial data. Table 7.2.2 shows the results of four different ANOVAs. Top left is the correct analysis for the unblocked design and bottom right is the correct analysis for the blocked design. The blocking has been very successful in lowering the residual mean square as there really were large differences

TABLE 7.2.1 Arrangement for Simulated Blocking Experiment

Plot	Yield	Arrangement		Plot	Yield	Arrangement	
		Blocked	Random			Blocked	Random
1	532	A	E	13	646	B	A
2	609	B	D	14	632	F	D
3	607	E	E	15	674	A	F
4	685	F	F	16	601	C	B
5	722	D	C	17	656	D	B
6	639	C	C	18	648	E	A
7	681	D	E	19	534	A	F
8	663	E	F	20	535	B	A
9	736	C	B	21	539	F	A
10	727	A	C	22	604	C	D
11	777	B	E	23	496	E	B
12	634	F	C	24	543	D	D

TABLE 7.2.2 ANOVA for Simulated Blocking

Source of Variation	D.F.	Random Arrangement		Blocked Arrangement	
		S.S.	M.S.	S.S.	M.S.
Unblocked Analysis					
UNITS STRATUM					
TREATMNT	5	19602	3920	8285	1657
RESIDUAL	18	105328	5852	116645	6480
TOTAL	23	124930	5432	124930	5432
Blocked Analysis					
BLOCKS STRATUM	3	15801	5267	79596	26532
BLOCKS.*UNITS* STRATUM					
TREATMNT	5	19602	3920	8285	1657
RESIDUAL	15	89527	5968	37050	2470
TOTAL	20	109130	5456	45335	2267
GRAND TOTAL	23	124930		124930	

between the blocks. As there is no treatment effect the treatment mean square is an estimate of the residual variance. Consequently the treatment mean square is lower in the blocked design. Any real treatment effect would increase the treatment sums of squares equally in both experiments, but the lower residual mean square would mean that the blocked experiment would be more likely to produce a significant F-ratio.

The incorrect analyses are also interesting. Ignoring the blocking, as in the top right ANOVA, gives a larger residual mean square than the unblocked experiment. The reason is that treatment mean squares is the same low value as for the blocked experiment without the compensating removal of the block sum of squares. The total is the same, so the residual must be larger. The practical consequence of this is that if an experimenter cheats by arranging the treatments nonrandomly, and if this cheating successfully lowers the variability of the treatment parameter estimates, the estimate of that variability will be increased.

The final incorrect analysis shows what sometimes happens when data from a random experiment is presented in a table. Because the rows in a table look temptingly like blocks, a block factor is included in the analysis. In this case blocks were not quite random because the table was filled using plots from left to right across the experiment. Even so the block effect is smaller than the residual, and five degrees of freedom are lost from the residual where they usefully belong.

7.2.2 Complete Randomized Block - Theory

When a treatment effect is added to (7.2.2) the linear model becomes

$$y_{ij} = \mu + \beta_j + \tau_i + \varepsilon_{ij} \qquad 1 \leq i \leq k,\ 1 \leq j \leq r \qquad (7.2.3)$$

$$X^T X = \left|\begin{array}{c:c:c} n & k\,\mathbf{1}^T & r\,\mathbf{1}^T \\ \hdashline k\,\mathbf{1} & k\,I & N^T \\ \hdashline r\,\mathbf{1} & N & r\,I \end{array}\right|$$

where N is a k×r matrix of 1's. Note that the (i,j)-th element of N is the number of times the i-th treatment appears in the j-th block. N is called the incidence matrix and any block design is completely described by it.

Formally the introduction of blocks to a model bears some resemblance to the splitting of the treatment effect into factors α and β in (6.4.1). From (6.4.1)

$$y_{ij} = \mu + \alpha_l + \beta_m + (\alpha\beta)_{lm} + \varepsilon_{ij} \qquad 1 \leq i \leq uv,\ 1 \leq j \leq r$$

Compare this with (7.2.3)

$$y_{ij} = \mu + \beta_j + \tau_i + \varepsilon_{ij} \qquad 1 \leq i \leq k,\ 1 \leq j \leq r$$

The first can be relabelled with l and m replacing i:

$$y_{lmj} = \mu + \alpha_l + \beta_m + (\alpha\beta)_{lm} + \varepsilon_{lmj} \qquad 1 \leq l \leq u,\ 1 \leq m \leq v,\ 1 \leq j \leq r$$

and the second with l and m replacing i and j respectively

$$y_{lm} = \mu + \beta_m + \tau_l + \varepsilon_{lm}$$

The similarity arises because we have treated the complete randomized block like a factorial experiment with one factor being blocks and the other treatments. The results (6.4.7) can therefore be used to calculate the sums of squares.

The degrees of freedom which in a factorial experiment are used to estimate an interaction are used in a block experiment to estimate the residual variance. Blocks represent a grouping of the experimental units before treatments are applied. They are not something to be tested or estimated. Therefore in the block model we have described any conclusions from the experiment apply to all the experimental units, and any differences in treatment responses from one group to another must be regarded as random variation. If for example the groups are blocks of land which are of clearly different soil types, or are of animals of different breeds, the experimenter may wish to draw separate conclusions for each group and our model would not apply. In such cases groups become a factor of interest for their own sake, which raises the possibility of an interaction which can only be eliminated by assuming it to be zero. Unless this assumption is reasonable, no estimate of the residual variance can be obtained.

7.3 BALANCED INCOMPLETE BLOCK DESIGNS

Not every experimenter is fortunate enough to be able to group experimental units into blocks of just the right size to contain a complete set of treatments. Sometimes there are a large number of treatments to be compared, as in plant breeding experiments screening large numbers of new varieties. Sometimes experimental units occur naturally in very small groups, identical twin animals for example. If treatments are equally important, and if any one pairwise comparison is of as much interest as any other, then a balanced incomplete block design is likely to be appropriate. A design is said to be balanced if every treatment difference is estimated with the same accuracy, and it is achieved in an incomplete block design by ensuring that each treatment appears in the same block with each other treatment equally often.

Example 7.3.1 Incomplete block example

Suppose a factory receives raw material in batches large enough for three runs so that blocks of three runs are appropriate. Now if six treatments are to be compared the design in Figure 7.3.1 can be used. There are ten batches (batch = block), each of three runs, and six treatments each replicated five times. Each pair of treatments appears together in the same block twice, so that the design is balanced.

Unfortunately, only a few combinations of block size, replication and number treatments can form a balanced incomplete block design. If there are k treatments and if each block must contain q experimental units, a design is always possible by taking every selection of q units from k treatments. The number of ways of

	Batches									
Treatments	1	2	3	4	5	6	7	8	9	10
A	x			x		x	x	x		
B		x			x		x	x	x	
C			x			x		x	x	x
D	x	x	x				x			x
E	x		x	x	x				x	
F		x		x	x	x				x

FIGURE 7.3.1 Six treatments (A-F) with batches as blocks.

selecting q objects from k, ${}_kC_q$, is large, usually too large for a practical experiment. Finding smaller designs is an interesting mathematical exercise, but we will rely on tables of possible designs, one of which is in Fisher and Yates, 1974. There are however three straight-forward requirements. Suppose we have k treatments replicated r times in p blocks of q experimental units, and let λ be the number of times each pair of treatments appears together in the same block. Then

$$kr = pq = n \tag{7.3.1}$$

$$\lambda(k-1) = r(q-1) \tag{7.3.2}$$

$$p \geqq k \tag{7.3.3}$$

(7.3.1) needs no explanation. To see (7.3.2) consider a particular treatment, X say. This appears in r blocks, and in these r blocks are r(q-1) other experimental units. Also X is in the same block with each other treatment λ times, so there must be $\lambda(k-1)$ experimental units in blocks with X. Hence (7.3.2). (7.3.3) is not as obvious as its simplicity suggests, and we will have to leave it for you to take on trust.

7.3.1 Solution of the Normal Equations

The model for an incomplete block design is

$$y_{ij} = \mu + \beta_j + \tau_i + \varepsilon_{ij} \tag{7.3.4}$$

but only some combinations of i and j occur.

$$X^T X = \left| \begin{array}{c:c:c} n & q\,\mathbf{1}^T & r\,\mathbf{1}^T \\ \hdashline q\,\mathbf{1} & I\,q & N^T \\ \hdashline r\,\mathbf{1} & N & I\,r \end{array} \right| \tag{7.3.5}$$

where N is the incidence matrix, with elements

$$n_{ij} = 1 \text{ if the i-th treatment is in the j-th block}$$
$$= 0 \text{ otherwise}$$

From $X^T X$ the normal equations can be obtained:

$$y_{..} = n\,m + q \sum b_i + r \sum t_i \tag{7.3.6}$$

$$y_{i.} = r\,m + r\,t_i + \sum_j n_{ij}\,b_j \tag{7.3.7}$$

$$y_{.j} = q\,m + \sum_h n_{hj}\,t_h + q\,b_j \tag{7.3.8}$$

We will not reproduce the detail of solving these equations. (7.3.8) gives the b's which can then be substituted in (7.3.7) to find the t's. As usual the t's are restricted to add to zero. This leads to the following estimate for the τ_i's.

$$\lambda k t_i = q y_{i.} - \sum_j n_{ij}\,y_{.j} \tag{7.3.9}$$

= qr (mean of all units receiving i-th treatment
- mean of all units in blocks containing i-th treatment)

So each τ_i is estimated just from those blocks which contain the i-th treatment. The following useful results can be proved using the methods used to prove (5.4.5) and (5.4.6).

$$\text{var } t_i = r(q-1)q\,\sigma^2/(\lambda k)^2 \tag{7.3.10}$$

$$\text{cov}(t_i, t_h) = -q\,\sigma^2/(\lambda k)^2 \qquad i \neq h \tag{7.3.11}$$

$$\text{var}(t_i - t_h) = 2q\,\sigma^2/(\lambda k) \tag{7.3.12}$$

If g is a contrast

$$\text{var } g = (\sum c_i^2)q\,\sigma^2/(\lambda k) \tag{7.3.13}$$

It is interesting to compare the variance of a treatment difference from an incomplete block experiment with that from an unblocked experiment of the same size. The ratio of the variances is the efficiency of the design and is equivalent to the ratio of the number of experimental units required for each type of experiment to give equal precision.

$$\begin{aligned}\text{Efficiency of incomplete block} \\ &= (\text{variance from unblocked})/(\text{variance from incomplete block}) \\ &= (2/r)/(2q/\lambda k) \\ &= (\lambda k)/(qr) \end{aligned} \tag{7.3.14}$$

For Example 7.3.1 the efficiency is $(2\times6)/(5\times3)$ or 80%. But presumably the model fits better if batches can be accounted for in the design, and the variance of each run is lowered. If it can be lowered to less than 80% of what it would be ignoring blocks, (or using large complete blocks) the treatment comparisons will be more precise overall.

This lower efficiency arises because a treatment difference is estimated partly within blocks (two in Example 7.3.1), and partly between (four in Example 7.3.1) blocks. A comparison between blocks requires an estimate of the block difference, and it is this extra estimate which is not required in an unblocked experiment, nor in an experiment in complete blocks.

7.3.2 Analysis of Variance

A glance at the normal equations shows that blocks will not be orthogonal to treatments. As we are using model (7.2.2) where treatments are estimated within blocks, the treatment sum of squares must be

TABLE 7.3.1 ANOVA for Incomplete Block Design

Source	S.Sqs.	D.F.
Blocks (ignoring treatments)	$\sum q\bar{y}_{.j}^{2} - n\bar{y}_{..}^{2}$	p-1
Treatments (adjusted for blocks)	$\sum (mk)t_i^2/q$	k-1
Residual	Subtraction	n-k-p+1
Total	$\sum y_{ij}^2 - n\bar{y}_{..}^{2}$	n-1

determined as the increase in model sum of squares through adding a treatment factor to a model already including the block factor. In the nomenclature of Chapter 3, Section 3.4 this is the treatment sum of squares adjusted for blocks. We have:

$$SS(\mu) = n\,\bar{y}_{..}^{2} \qquad \text{from } y_{ij} = \mu + \varepsilon_{ij}$$

$$SS(\beta,\mu) = \sum_j q\,\bar{y}_{.j}^{2} - n\,\bar{y}_{..}^{2} \qquad \text{from } y_{ij} = \mu + \beta_j + \varepsilon_{ij}$$

$$SS(\tau,\beta,\mu) = \sum_i y_{ij}t_i + \sum_j y_{j.}\,b_j \quad \text{from } y_{ij} = \mu + \beta_j + \tau_i + \varepsilon_{ij}$$

$$SS(\tau, \text{adjusted for blocks}) = SS(\tau,\beta,\mu) - SS(\mu,\beta).$$

Some rather tedious algebra shows that this later quantity is

$$= (\lambda k) \sum t_i^2/q \qquad (7.3.15)$$

The analysis of variance table is therefore as in Table 7.3.1. The residual gives an estimate of the standard deviation to use in (7.3.10) to (7.3.13) as usual.

Example 7.3.3 Analysis of incomplete block experiment

We show the analysis of data from Example 7.3.1. The data itself is in Table 7.3.2 and the sums of squares are calculated as below.

Block sum of squares (ignoring treatments)

$$= \{81.7^2 + 102.4^2 + \ldots + 87.5^2\}/3 - 876.6^2/30$$
$$= 209.73$$

Treatment sum of squares (adjusted for blocks)

$$= \{(-16.7)^2 + (45.7)^2 + \ldots + (25.1)^2\}/\{2\times 6\times 3\} \quad \text{from (7.3.15)}$$
$$= 151.31$$

Total sum of squares

$$= \{25.6^2 + \ldots + 33.9^2\} - 876.6^2/30$$
$$= 395.07$$

The ANOVA table can now be completed as in Table 7.3.3 and it is possible to make all the inferences of Chapter 6 using the results (7.3.10) to (7.3.13). For example, the standard deviation of a treatment difference is

$$\text{s.d.}(t_i - t_j) = \sqrt{(2\times 3\times 2.269)/(2\times 6)} = 1.065 \quad \text{from (7.3.12)}$$

TABLE 7.3.2 Data From Incomplete Block Example

Batch	Treatment A	B	C	D	E	F	Total
1	25.6			26.2	29.9		81.7
2		35.2		31.0		36.2	102.4
3			34.2	30.3	34.1		98.6
4	25.0				26.3	27.5	78.8
5		32.5			29.7	29.5	91.7
6	25.3		27.7			28.6	81.6
7	25.9	29.6		21.4			76.9
8	27.9	33.4	25.5				86.8
9		34.0	26.4		30.2		90.6
10			28.5	25.1		33.9	87.5
TOTAL	129.7	164.7	142.3	134.0	150.2	155.7	876.6
$\sum_j n_{ij} y_{ij}$	405.8	448.4	445.1	447.1	441.4	442.0	
$\lambda k t_i$	-16.7	45.7	-18.2	-45.1	9.2	25.1	

TABLE 7.3.3 ANOVA Table for Incomplete Block Experiment

Source	S.Sqs.	DF	M.S.Sqs	F
Blocks	209.73	9		
Treatment	151.31	5	30.262	13.3
Residual	34.03	15	2.269	
Total	395.07	29		

7.3.3 Comparison Between Blocks

In a complete block design blocks are orthogonal to treatments, so that no information is lost by using (7.2.2) instead of (7.2.1). In an incomplete block design blocks are not orthogonal to treatments so that treating blocks as experimental units gives extra information for estimating treatments. The model is

$$B_j = q\mu + \sum_i n_{ij}\tau_i + (q\xi_j + \varepsilon_{.j}) \qquad (7.3.16)$$

where the B_j are block totals. Note that there have to be more blocks than treatments - (7.3.3) ensures that there will be at least as many. These estimates are called the inter-block estimates, in contrast to the intra-block estimates of the earlier section. As the two estimates are independent, a weighted average can be formed using the inverse of their variances. We will not pursue this topic, but we will be considering designs where blocks are experimental units in the next chapter. A more detailed treatment of general block designs may be found in John, 1971.

7.4 CONFOUNDING

When factorial experiments were first introduced in Section 6.4, we commented that they inevitably tended to have many treatments. They will therefore be difficult to fit into reasonable sized blocks, and are prime candidates for some form of incomplete block design. There is an essential incompatibility, however, between the requirements of a factorial experiment and the properties of the balanced incomplete block design. In a balanced design every treatment difference is estimated with the same precision. With a factorial design individual treatment differences are not usually important. It is differences between factor means, or main effects, which are important. Two factor interactions are of secondary importance and three factor or higher interactions are often of no interest at all. An ideal form of blocked experiment would be one which gives full information on the main effects, and which sacrifices information on the interactions.

A technique for doing this is confounding. Experimental design is intended to enable the causes of observed differences to be unambiguously determined. The polio vaccine experiment described in Chapter 5 was a bad experiment because differences in the incidence of polio had several possible causes besides the use of vaccine. Such causes are said to be confounded. However, if a high order interaction is not of any particular interest, nothing is lost if it is confounded with existing differences between the groups of experimental units. We will explain how this is done in the context of an experiment with many two level factors.

Example 7.4.1 Magnesium on potatoes

In a recent experiment to find out whether magnesium was necessary for potato crops four factors were included in a 2×2×2×2 experiment. They were phosphate (p), potash (k), nitrogen (n) and magnesium (m). In each case the two levels were "none" and "some", the "some" being chosen to be reasonable maximum rate. There were two replicates, set out as in Table 7.4.1. We will refer to this example in the following discussion, and the analysis will be shown in the next example.

A simplified notation is used for labelling the treatments. A treatment could be specified as p2 k1 n1 m2, meaning level 2 of phosphate and magnesium, and level 1 of potash and nitrogen. However

TABLE 7.4.1 Layout and Data for Potato Fertiliser Experiment

Plot	Ntrgn	Phos	Potsh	Mgnsm	Yield	Plot	Ntrgn	Phos	Potsh	Mgnsm	Yield
1	1	2	2	1	2850	17	2	1	1	2	2960
2	2	1	1	2	3070	18	1	2	1	2	2670
3	2	2	2	2	3420	19	1	2	2	1	3310
4	1	1	2	2	3110	20	2	2	1	1	3020
5	1	1	1	1	2810	21	2	1	2	1	2640
6	2	2	1	1	3110	22	2	2	2	2	2890
7	2	1	2	1	2940	23	1	1	2	2	3110
8	1	2	1	2	3190	24	1	1	1	1	3100
9	1	1	2	1	2860	25	2	1	2	2	2720
10	1	2	1	1	2850	26	2	1	1	1	2920
11	1	2	2	2	2300	27	2	2	2	1	3110
12	1	1	1	2	2880	28	1	2	2	2	3370
13	2	1	1	1	3220	29	1	2	1	1	2700
14	2	1	2	2	3580	30	1	1	2	1	3280
15	2	2	1	2	3100	31	2	2	1	2	3300
16	2	2	2	1	3680	32	1	1	1	2	2850

since there are only two levels the presence or absence of the letter is sufficient. p2 k1 n1 m2 is written pm, p2 k2 n2 m2 is pknm, and by convention, p1 k1 n1 m1 is (1). Depending on the context, these letters also stand for the mean of all observations on that treatment. Then, capital letters are used for the effects. K stands for the effect of potash, and MK for the interaction between magnesium and potash. The effects themselves were defined in Chapter 6, Section 4. The K effect is defined as

$$\begin{aligned} K &= \{[\text{mean of all treatments including } k] \\ &\quad - [\text{mean of all treatments excluding } k]\}/2 \\ &= \{[k + kp + km + kmp + kn + kpn + kmn + kmpn] \\ &\quad - [(1) + p + m + mp + n + pn + mn + nmp]\}/16 \end{aligned} \qquad (7.4.1)$$

An interaction is a measure of how the effect of one factor changes depending on the level of another, so that, from Example 6.2.3, the KM effect is

KM = {[K effect in the presence of m]
 - [K effect in the absence of m]}/2

 = {[(mean of all treatments including k and m)
 - (mean of all treatments including m, not k)]/2
 - [(mean of all treatments including k, not m)
 - (mean of all treatments including neither k nor m)]/2}/2

 = {[(km + nkm + pkm + pnkm) - (m + pm + nm + pnm)] (7.4.2)
 - [(k + kn + pk + pkn) - ((1) + p + n + pn)]}/16

Extending this principle, a three factor interaction, NKM say, is defined as

{[KM interaction effect in the presence of n] (7.4.3)
 - [KM interaction effect in the absence of n]}/2

and the four factor interaction PNKM is defined as

{[NKM interaction effect in the presence of p]
 - [NKM interaction effect in the absence of p]}/2

Interactions of any order can be defined by continuing this procedure.

Note that all of these contrasts are the differences between two groups of eight treatment means. We will call one the + group, the other the - group. Now consider the KM interaction effect. The members of the + group are those in the first set of round brackets, which are those treatment combinations containing both m and k, and those in the last set of round brackets, which are those combinations containing neither m nor k. So the + group is all treatment combinations with either 0 or 2 of m or k. Similarly the - group is all treatment combinations with 1 of m or k.

By induction the general rule can be proved: (7.4.4)

Divide the treatments into two groups - those with an even number of letters in common with the interaction effect in one group and those with an odd number of letters in common in the other. The + group is the one which includes the treatment combination with all the letters of the interaction effect.

Thus the NKM interaction is a comparison between the groups:

+: n, k, m, nkm, pn, pk, pm, pnkm
-: nk, km, nm, (1), pnk, pkm, pnm, p

and the PNKM interaction is a comparison between the groups:

+: pn, pk, pm, nk, nm, km, pnkm, (1)
-: p, n, k, m, nkm, pkm, pnm, pnk

We can now explain how confounding was used to permit blocks of eight in Example 7.4.1. Block 1 (plots 1 - 8) and Block 3 (plots 17 - 24) comprise the + group of the PNKM interaction, and Block 2 (plots 9 - 16) and Block 4 (plots 25 - 32) comprise the - group. This means that any differences between these blocks of plots will cause the same observable differences as would a PNKM interaction. As this four factor interaction was of no interest (and almost certainly zero) nothing of importance was lost. The residual variance was lowered because blocks of land large enough for eight plots were more uniform than blocks of land large enough for sixteen plots.

The procedure for confounding is to decide first on a comparison which is of no interest, then to use (7.4.4) to determine the groups of treatments involved in the comparison, and finally to allocate each of these groups within a block.

Example 7.4.2 Analysis of potato experiment

A GENSTAT analysis of the results is given in Table 7.4.2 . Note that SS(PKNM) appears in the BLOCK stratum, rather than the BLOCK. *UNITS* stratum. GENSTAT works out for itself what is confounded and does the correct analysis. A human would calculate the TOTAL sum of squares for blocks from the four block totals using (6.4.7). The split into components is not really required, as SS(PKNM) is automatically included in it. The various sums of squares for treatment factors are calculated separately using (6.4.7) and (6.4.8).

There is no need to stop at one confounded effect. Because effects are orthogonal any additional confounded effect splits each existing group of plots into two equal sized groups, so that the

TABLE 7.4.2 ANOVA for Potato Fertiliser Experiment

***** ANALYSIS OF VARIANCE *****

VARIATE: Yield

SOURCE OF VARIATION	DF	SS	MS	VR
Blocks STRATUM				
Ntrgn.Phos.Mgnsm.Potsh	1	8450	8450	0.393
RESIDUAL	2	43025	21513	
TOTAL	3	51475	17158	
Blocks.*UNITS* STRATUM				
Ntrgn	1	186050	186050	1.528
Phos	1	21013	21013	0.173
Mgnsm	1	450	450	0.004
Potsh	1	63013	63013	0.517
Ntrgn.Phos	1	171113	171113	1.405
Ntrgn.Mgnsm	1	14450	14450	0.119
Phos.Mgnsm	1	25313	25313	0.208
Ntrgn.Potsh	1	23113	23113	0.190
Phos.Potsh	1	9800	9800	0.080
Mgnsm.Potsh	1	6613	6613	0.054
Ntrgn.Phos.Mgnsm	1	17113	17113	0.140
Ntrgn.Phos.Potsh	1	42050	42050	0.345
Ntrgn.Mgnsm.Potsh	1	12013	12013	0.099
Phos.Mgnsm.Potsh	1	217800	217800	1.788
RESIDUAL	14	1705175	121798	
TOTAL	28	2515075	89824	
GRAND TOTAL	31	2566550		

number of blocks in a rep must be a power of two. Since the block degrees of freedom must equal the number of confounded effects the number of confounded effects must be one less than a power of two. To see how this fits in with the odd and even numbers of letters in common with the confounded effect, consider what would happen if we decided to confound the PKM effect in the previous example. Being a three factor interaction we may be happy to lose information about it. For PKM, the + group is $\underline{p}$, $\underline{k}$, $\underline{m}$, pn, kn, mn, $\underline{pkm}$, pkmn and the - group is km, $\underline{kmn}$, mp, $\underline{mpn}$, pk, $\underline{pkn}$, (1), $\underline{n}$ where the underlined treatments are the - group from confounding PKNM. With both PKM and PKMN confounded the four blocks are as shown in Table 7.4.3. Unfortunately the N main effect is now confounded between Blocks 1 and 4

TABLE 7.4.3 Two Factors Confounded

		PKNM confounded + group	PKNM confounded - group
PKM	+ group	pn kn mn pkmn Block 1	p k m pkm Block 3
confounded	- group	km mp pk (1) Block 2	kmn mpn pkn n Block 4

and Blocks 2 and 3. The three degrees of freedom for blocks correspond to the N, PKM and PKMN effects.

There is a general rule for discovering the third confounded effect given two:

(7.4.5)

> If any two effects are taken and their common letters struck out, the remaining letters determine the third confounded effect.

So taking PKM and PKMN, striking out the common letters P, K and M leaves N, the third confounded effect. Using this rule makes it easy to see that there is no way of arranging the treatments into four blocks without at least confounding two factor interactions.

Any two effects determine four sets of individual treatments as in Table 7.4.3. The rule (7.4.5) just stated also provides a rule for finding the third orthogonal effect for comparisons between that set of four treatment means. Algebraists will note that the combination of two effects to give the third is a form of multiplication where $x^2 = 1$. For example

$$(P \times K \times M) \times (P \times K \times M \times N) = P^2 \times K^2 \times M^2 \times N = N$$

Under this operation the set of all effects form an algebraic group with the identity (I), and the set of all confounded effects together with (I) form a subgroup.

7.4.1 Negligible Contrasts

If the effect measured by a contrast is known to be negligible, the sum of squares for that contrast will be a measure of the residual

TABLE 7.4.4 Combining High Order Interactions

***** ANALYSIS OF VARIANCE *****

SOURCE OF VARIATION	DF	SS	DF	SS	MS	VR
BLOCKS STRATUM						
Phos.Potass.Boron	1	82220				
Lime.Boron.Magnes	1	3530	3	93560	31187	
Lime.Phos.Potass.Magnes	1	7810				
TOTAL	3	93560	3	93560	31187	
BLOCKS.*UNITS* STRATUM						
Lime	1	30500	1	30500	30500	14.4
Phos	1	44550	1	44550	44550	21.0
Potass	1	7260	1	7260	7260	3.4
Boron	1	0	1	0	0	0.0
Magnes	1	6550	1	6550	6550	3.1
Lime.Phos	1	880	1	880	880	0.4
Lime.Potass	1	3120	1	3120	3120	1.5
Phos.Potass	1	3830	1	3830	3830	1.8
Lime.Boron	1	15310	1	15310	15310	7.2
Phos.Boron	1	30	1	30	30	0.0
Potass.Boron	1	10	1	10	10	0.0
Lime.Magnes	1	1110	1	1110	1110	0.5
Phos.Magnes	1	20	1	20	20	0.0
Potass.Magnes	1	10	1	10	10	0.0
Boron.Magnes	1	410	1	410	410	0.2
Lime.Phos.Potass	1	240				
Lime.Phos.Boron	1	10080				
Lime.Potass.Boron	1	30				
Lime.Phos.Magnes	1	1200				
Lime.Potass.Magnes	1	40				
Phos.Potass.Magnes	1	0				
Phos.Boron.Magnes	1	210	13	27600	2123	
Potass.Boron.Magnes	1	2080				
Lime.Phos.Potass.Boron	1	2110				
Lime.Phos.Boron.Magnes	1	20				
Lime.Potass.Boron.Magnes	1	3610				
Phos.Potass.Boron.Magnes	1	7750				
Lime.Phos.Potass.Boron.Magnes	1	220				
TOTAL	28	141180				
GRAND TOTAL	31	234740				

***** TABLES OF MEANS *****

Lime	NONE	2000kg
	633.1	571.3

Phos	NONE	80kg
	564.9	639.5

Potass	NONE	140kg
	617.3	587.1

Boron	NONE	2.4kg
	602.2	602.2

Magnes	NONE	25kg
	587.9	616.5

Boron / Lime	NONE	2.4kg
NONE	655.1	611.0
2000kg	549.6	593.0

variance. An experiment with several factors can therefore use high order interactions to give an estimate of the residual variance. It therefore need not be replicated.

Example 7.4.3 Single replicate experiment

The output in Table 7.4.4 is the analysis of a single replicate, five factor fertilizer experiment. Two separate ANOVA tables have been put together. The left hand one shows the complete breakdown into individual effects, and the right hand one shows all the three factor interactions combined into the residual, except for those confounded with blocks. Main effect means are also shown, together with the Lime × Boron interaction, which was shown to be significant. Note that the confounded effects satisfy (7.4.5).

7.5 MISCELLANEOUS TRICKS

We finish this chapter by mentioning three techniques which are extensions of the principles already covered.

7.5.1 Fractional Replication

We have already seen that it is possible to use higher order interactions to give an estimate of the residual variance so that the complete set of treatments need not be replicated. It is further possible to run an experiment with just a half or a quarter of the complete set. We will just give an example and not deal with the topic fully.

Example 7.5.1 Concrete strength

In industrial experiments each individual run is likely to be expensive, in contrast to agricultural experiments where the cost of the experiment will not depend strongly on the number of plots. Also machines tend to give less variable results than plants so the replication need not be high. Suppose that an experiment is required to assess the effect on concrete strength of the five factors listed in Table 7.5.1. What can be done with just 16 batches of concrete? First consider a single complete replicate of 32 batches divided into

TABLE 7.5.1 Factors in Concrete Experiment

Factor	Levels
A. Cement content	Normal,low
B. Proportion of sand	Normal,low
C. Moisture content	High,low
D. Curing time	One day,one week
E. Curing temperature	10°C,20°C

two by confounding ABCDE. Then the A effect also divides the 32 batches into two and we can construct Table 7.5.2, just like Table 7.4.3, which shows the 32 batches divided into four groups. By (7.4.5) the effect BCDE is also a comparison between these four groups. We have:

$$\begin{aligned} A &= \{(I) - (II) + (III) - (IV)\}/16 \\ BCDE &= \{(I) - (II) - (III) + (IV)\}/16 \\ A + BCDE &= \{(I) - (II)\}/8 \end{aligned}$$

If only the 16 batches corresponding to groups (I) and (II) were completed the A effect would equal the BCDE effect and column of the X matrix for the A effect would be exactly the same as the column for the BCDE effect. The A effect and the BCDE effect are said to be aliased. The useful result is that their sum can be estimated. Combine this with the knowledge that the four factor interaction could almost certainly be assumed zero and we have a way of estimating the A effect. It is the difference between (I) and (II).

Similarly, all the other effects can all be arranged in pairs of aliases whose sum can be estimated. For the concrete experiment the aliases are

Main effects: A and BCDE, B and ACDE, C and ABDE, D and ABCE, E and ABCD
Interactions: AB and CDE, AC and BDE, AD and BCE, AE and BCD, BC and ADE, BD and ACE, BE and ACD, CD and ABE, CE and ABD, DE and ABC.

These account for all 15 degrees of freedom from the 16 batches, leaving nothing for residual. Consequently this may not seem to be a

TABLE 7.5.2 Division of Treatments

	A effect + group		A effect − group	
Block 1: + group for ABCDE	I:	a abc acd abcde abd abe ace ade	II:	bde bcd b c cde bce d e
Block 2: − group for ABCDE	III:	ab ac ad ae abde abcd acde abce	IV:	cd ce de (1) bc bd be bcde

particularly useful experiment, particularly since there are likely to be strong two factor interactions - they cannot be used to estimate the residual variance. However it may be possible to use an estimate of residual from a previous experiment, or possibly a few of the two factor interactions may be known to be zero. Alternatively, if all the effects were zero their estimates would have identical independent normal distributions. The normal plot described in Chapter 6, Section 5.1 could therefore show especially large effects in the same way as it showed large differences between means.

Whatever technique is used, where the residual variance is small this type of experiment can give useful information most economically.

7.5.2 Partial Confounding

So far we have looked at the balanced incomplete block experiment, which loses information uniformly over every contrast, and the confounded factorial experiment, which loses information completely on some effects. If an experiment has more than one replicate it is possible to confound a different effect in each replicate. At least limited information can then be obtained on all the effects. To analyse such an experiment the block sum of squares is found from the block means as usual, and the treatment sum of squares for unconfounded effects is found from the treatment means over all replicates, again as usual. Sums of squares for and estimates of confounded effects are found from the treatment means in the replicates where the effect is not confounded. Note that the partially confounded effects are part of the block by treatment interaction because they represent the different treatment effects in each block. Their degrees of freedom therefore reduce the residual degrees of freedom.

7.5.3 More Than Two Levels

To extend two level factors to more than two levels requires more development of the algebraic approach mentioned briefly at the end of Section 7.4. Of course with more than two levels the experiment becomes very large if there are three or more factors and, as we saw in Example 7.5.1, fractional replication can give little information on even two factor interactions unless there are more than four factors. Another practical problem is that because extreme levels of two or more factors may not represent a realistic treatment combination, a complete factorial may mean including treatments of little practical interest. Designing efficient combinations of treatments for multifactor multilevel experiments is response surface methodology, a complete subject in itself. An introduction to the theory of response surface designs is given in John, 1971.

However, a useful trick for a four level factor is to code it as two two level factors. That is, the four levels 1, 2, 3, 4 are coded (1), a, b, ab. This enables an experiment involving a four level factor, perhaps combined with some two level factors, to be treated as a multifactor two level experiment. All the techniques of confounding described in Section 7.4 can then be used with the caution that the AB effect is really part of a main effect.

PROBLEMS

7.1. The experiment to study weedicides on a pea crop (Appendix C 7) was in fact a complete randomized block.

(i) Complete the analysis of variance including the sums of squares for the blocks, treatments, and residual.
(ii) Did blocking improve the precision of the experiment?

7.2. As part of a study on the nutritional quality of oats six varieties of oat kernel with their hulls removed are subjected to a mineral analysis. The following measurements of protein by percent dry weight are recorded.

Variety	1	2	3	4	5	6	Total
1A	19.1	16.3	16.3	17.5	16.2	21.1	106.5
1B	20.3	17.9	18.2	18.0	16.9	21.4	112.7
1C	20.3	16.9	17.4	17.6	15.9	21.4	109.5
2A	19.6	17.6	17.5	17.6	14.8	20.5	107.6
2B	18.6	16.7	16.3	17.4	16.0	21.1	106.1
2C	20.1	17.3	17.9	18.0	16.7	21.6	111.6
Total	118.0	102.7	103.6	106.1	96.5	127.1	654.0

Calculate an ANOVA table for the data for the following cases:

(i) The experiment was planted as two replicates and the A,B,C figures refer to the three samples taken from each plot for protein determinations.
(ii) The experiment was planted as six replicates and so was a straightforward randomized block.
(iii) A convenient block size was twelve plots, so two of the treatments were randomized over each of three blocks, the blocks being A, B, C.

Does the data suggest which of the above designs might in fact have been used? For (iii) calculate and plot residuals. Does the model appear adequate?

7.3. Fifteen pairs of identical twin calves were used in an experiment to compare threee worm drenchs. Their weight gains (kg) over the course of the experiment were:

Twin Pair	1	2	3	4	5	6	7	8	9	10	11	12	13	14	15
Drench A	60	70	47	45	35	50	40	51	50	50					
Drench B	65	56	44	57	56						71	59	65	37	65
Drench C						40	37	40	42	52	55	35	47	21	56

(i) Are there any differences between the drenches?
(ii) How much precision was gained by using identical twins?

7.4. The data below are from a classical experiment conducted at Rothamstead to compare the following fertilizers in all combinations:

Sulphate of ammonia	None or 6 cwt/acre	(s)
Superphosphate	None or 0.5 cwt/acre	(p)
Potash	None or 1.0 cwt/acre	(k)
Salt	None or 5 cwt/acre	(n)
Dung	None or 10 ton/acre	(d)

The layout of the experiment and the plot yields are given in the diagram below.

pkd	nd	sk	spknd	d	pknd	k	snd
844	1104	1156	1508	1248	1100	784	1376
spn	kn	sd	p	spkd	skn	sp	pn
1312	1000	1176	888	1356	1376	1008	964
kd	spd	pnd	pkn	skd	spkn	knd	spnd
896	1284	996	860	1328	1292	1008	1324
sn	spk	(1)	sknd	pd	pk	n	s
1184	984	740	1468	1008	692	780	1108

The ANOVA table below gives the value of the sums of squares for all 31 treatment factors and interactions

(i) Determine what is confounded.
(ii) Construct the ANOVA table and form tables of means for all significant interactions. All three and four factor interactions are to be assigned to residual sum of squares.

***** ANALYSIS OF VARIANCE *****

SOURCE OF VARIATION	DF	SS
UNITS STRATUM		
S	1	887100
P	1	3040
K	1	720
N	1	144700
D	1	262100
S.P	1	380
S.K	1	48050
P.K	1	6270
S.N	1	16560
P.N	1	5830
K.N	1	30750
S.D	1	290
P.D	1	10
K.D	1	880
N.D	1	13780
S.P.K	1	290
S.P.N	1	290
S.K.N	1	10950
P.K.N	1	10
S.P.D	1	18050
S.K.D	1	18820
P.K.D	1	48670
S.N.D	1	240
P.N.D	1	970
K.N.D	1	800
S.P.K.N	1	340
S.P.K.D	1	13450
S.P.N.D	1	17670
S.K.N.D	1	3870
P.K.N.D	1	5200
S.P.K.N.D	1	1460
TOTAL	31	1561500

7.5. Example 6.4.1 was in fact a simplified version of the full experiment. The full experiment had four factors and would be described as a $4 \times 2 \times 2 \times 2$ factorial experiment. There were no blocks and only one replicate. In each case the first level was the ideal, and the other levels were common bad practices. The results are as given below. After it an ANOVA table is given showing all sums of squares.

Exposure	Method of filling	Storage temp °C	Container Type			
			Tins	Jars	Clear	Opaque
Not exposed to light	From top	5	0.00	0.005	0.005	0.00
		30	0.05	0.06	0.04	0.03
	From bottom	5	0.00	0.015	0.01	0.01
		30	0.04	0.08	0.07	0.05
Exposed to light	From top	5	0.00	0.03	0.11	0.05
		30	0.05	0.07	0.08	0.07
	From bottom	5	0.00	0.10	0.12	0.06
		30	0.05	0.13	0.17	0.09

***** ANALYSIS OF VARIANCE *****

SOURCE OF VARIATION	DF	SS
UNITS STRATUM		
Cntnr	3	0.01191
Filling	1	0.00372
Temp	1	0.01182
Light	1	0.01598
Cntnr.Filling	3	0.00222
Cntnr.Temp	3	0.00065
Filling.Temp	1	0.00041
Cntnr.Light	3	0.00745
Filling.Light	1	0.00096
Temp.Light	1	0.00057
Cntnr.Filling.Temp	3	0.00103
Cntnr.Filling.Light	3	0.00083
Cntnr.Temp.Light	3	0.00051
Filling.Temp.Light	1	0.00006
Cntnr.Filling.Temp.Light	3	0.00038
TOTAL	31	0.05851

(i) Complete the ANOVA table using three and four factor interactions to give an estimate of the residual. Test all main effects and two factor interactions for significance.

(ii) Calculate tables of means for the main effects and interactions. Give a possible explanation for any significant interactions.

(iii) Design an experiment in two blocks of 16 samples which could be used if it were not possible to arrange 32 stored samples under uniform conditions. Treat the container type factor as two 2 level factors.

7.6. The results of an industrial experiment to measure the effect on % yield of a range of temperatures and pressures is given below. There was one replicate and two blocks, the two blocks being denoted by X or Y in the table.

(i) Show that if temperature levels are coded 1,a,b,ab and pressure levels are coded 1,c,d,cd the ABCD effect has been confounded.
(ii) Show that the ABCD effect is equivalent to the interaction between the quadratic effect of temperature and the quadratic effect of pressure.
(iii) Calculate an ANOVA table by extracting the two main effect sum of squares, the linear × linear and linear × quadratic components of interaction sum of squares, and using the interaction components involving cubic contrasts as a measure of the residual.
(iv) Hence test the various effects for significance.

Yield per cent

Pressure (kPa):		30	40	50	60
Temp.	20°	10.1 X	15.7 Y	18.5 Y	18.5 X
	30°	12.3 Y	17.2 X	20.6 X	19.8 Y
	40°	14.4 Y	18.5 X	21.0 X	19.3 Y
	50°	14.3 X	17.4 Y	20.9 Y	18.9 X

8
EXTENSIONS TO THE MODEL

8.1 INTRODUCTION

In this final chapter we look at three extensions to the type of design model or analysis described in earlier chapters. The first, having a hierarchy of experimental units, has already been hinted at in Chapter 7, Section 3.3. The second idea is of covariance. Here the X matrix has one (or more) columns of x's which are observations of a variable. Of course the model is still a standard linear model, but the simplicity of an X matrix which is all 1's and 0's is lost. Finally we look at an example of an experiment where the data is not balanced. Again this is a standard linear model, but nonorthogonality creates new problems in the analysis.

8.2 HIERARCHIC DESIGNS

In Chapter 7, Section 3.3 we introduced the idea of estimates being made either within groups, or between groups. We did not dwell on the between groups estimates in that chapter, but now we will consider ways of allotting treatments to experimental units in such a way that one treatment factor is estimated within groups and the other between groups. An example of this type of design is given in

Chapter 5, Problem 2(ii), the experiment in a controlled climate room. Pruning treatments were randomly applied to pots, watering treatments were randomly applied to groups of eight pots, and the humidity treatment was applied to rooms, and was unreplicated.

For the pruning treatment, pots were experimental units, and groups of eight pots were blocks. For the watering treatment, groups of eight pots were experimental units, and rooms were blocks. Since the humidity treatment was unreplicated no test of the humidity effect is possible. Rooms could be treated as complete blocks providing two replicates of the pruning and watering experiment, but as explained at the end of Chapter 7, Section 2.2 it is neccessary to assume that humidity has no interaction with other treatments.

We call groups of eight pots, <u>main plots</u> and single pots, <u>sub-plots</u>. Following (7.2.2) the model for the experimental unit grouping is, if y_{hij} is the observation on the h-th pot in the i-th group of eight pots in the j-th room

$$y_{hij} = \mu + \beta_j + \xi_{ij} + \varepsilon_{hij} \qquad 1 \leq j \leq 2, \quad 1 \leq i \leq 3, \quad 1 \leq h \leq 4 \qquad (8.2.1)$$

Strictly speaking the experimental unit structure should be described using a model with three levels, similar to (7.2.1), but with an extra term. Such a structure would need three linear models to encompass it. However we will only consider cases where the j indexes complete replicates of the experiment, so that comparisons between the largest groups contain no information about the treatments.

8.2.1. Split Plot Experiments

Experiments like that just described are called <u>split plots</u>. The traditional split plot would be a crop experiment with varieties randomized over large main plots which are split into sub-plots upon which fertilisers are randomized. Varieties have to be randomized over large plots because the machinery which sows the seed requires a large area to work in. Fertiliser on the other hand can be applied to quite small plots. With industrial processes temperature can be difficult to change because it can take a long time for a new temperature to become established throughout the process. On the other hand, flow rate can frequently be changed just by turning a tap. It is therefore convenient to establish a temperature and then run through all the flow rates in random order before establishing the next temperature. This again is a split plot. Their main advantage is a practical one. If a treatment factor is difficult to apply to a small experimental unit, where "small" may mean "small area" or "short time", and if this treatment is combined with a second factor which can be applied to a small experimental unit, a split plot design provides an economical use of resources.

Example 8.2.1 Crop sowing and harvesting

This experiment was one in a series of energy related studies. The crop was rape, whose seeds can provide vegetable oil, and the comparison was between four different sowing methods. Direct drilling places the seed and fertiliser directly into uncultivated ground, thereby saving a considerable amount of tractor work, and fuel. The experiment compared direct drilling with other more conventional cultivation methods. Machinery for doing this is large and bulky, and can only be handled on large plots. At harvest time, however, smaller machinery designed especially for experimental work can be used, so that the second factor, harvesting method, could be compared on sub-plots within the main plots. A sketch plan of the experiment is given in Figure 8.2.1. Data for the oil yields from a sample of 50 plants per plot are given in Table 8.2.1. The ANOVA table is shown in Table 8.2.2. The sums of squares for treatment factors are calculated, according to (6.4.7), and the total sum of squares for the main plot stratum is calculated from the main plot means also according to (6.4.7). Note that each factor is estimated in the stratum within which it was randomly allocated. 'Cultn' was randomly allocated to main plots within 'Reps', and 'Harvest' was randomly allocated to sub-plots within main plots. The 'Drct_cut' treatment is one where the crop is cut and left to dry. Birds are fond of this

Main Plot:	A	C	B	D	D	A	C	B	B	D	C	A	A	C	B	D
Subplots	b	b	a	c	b	c	b	b	a	c	b	c	a	b	a	a
	a	a	b	a	c	b	a	c	b	b	a	b	c	a	c	b
	c	c	c	b	a	a	c	a	c	a	c	a	b	c	b	c

FIGURE 8.2.1 Layout of split plot experiment.
A: Aitchison drill, B: Direct-drilled, C: Two-pass, D: Roller drilled
a: Windrowed, b: Desiccated, c: Direct cut

TABLE 8.2.1 Oil Yields of Cultivation Experimant

Cultn	Harvest	Blocks 1	2	3	4
Aitchisn	Windrwd	4167	4205	5104	4271
	Dessctd	3018	2328	2413	3948
	Drct_cut	2252	1056	1862	1187
Dirct_dl	Windrwd	2502	3391	3413	3544
	Dessctd	1417	2097	1796	2320
	Drct_cut	1978	233	86	250
2_pass	Windrwd	3685	4010	5050	3796
	Dessctd	3220	2627	3074	2169
	Drct_cut	2409	1931	1873	2024
Roll_drl	Windrwd	3457	2919	4962	4759
	Dessctd	2250	1884	4224	3560
	Drct_cut	2389	2865	1303	1199

TABLE 8.2.2 ANOVA For Cultivation Experiment

***** ANALYSIS OF VARIANCE *****

VARIATE: Yield

SOURCE OF VARIATION	DF	SS	MS	VR
Blocks STRATUM	3	1340175	446725	
Blocks.M_plot STRATUM				
Cultn	3	10223870	3407957	14.881
RESIDUAL	9	2061157	229017	
TOTAL	12	12285026	1023752	
Blocks.M_plot.S_plot STRATUM				
Harvest	2	46054728	23027364	39.458
Cultn.Harvest	6	1048711	174785	0.299
RESIDUAL	24	14006290	583595	
TOTAL	32	61109728	1909679	
GRAND TOTAL	47	74734928		

method of harvesting, and in this experiment they took a particular liking to reps 3 and 4. Consequently the RESIDUAL in the 'Blocks.-M_plot.S_plot' stratum is higher than in the 'Reps.M_plot' stratum. Usually the residual is smaller within sub-plots.

A property of the split plot experiment is that the main plot comparisons are normally less precise than the sub-plot comparisons, and also less precise than they would be in an ordinary randomized block with the same replication. The reason is that in an ordinary randomized block each main effect has extra replication provided by the other factors, but in a split plot the sub-plot treatment provides no extra replication of the main effects. If Example 8.2.1 had been a randomized block, the drilling main effect would have had twelve replications, but to achieve this the each plot would have had to be the same size as the main plots and the experiment would have had to be three times as large.

There is a small benefit to the sub-plot treatments through having the sub-plot treatments compared within small blocks. Note that a split plot experiment is equivalent to a factorial experiment with a main effect confounded, so that a split plot can sometimes be appropriate if the primary purpose of the experiment is to compare sub-plot treatments. For example, the classical varieties by fertilizer experiment may involve varieties whose properties are well known, and the intention of the experimenter might be to find the optimum fertilizer for each. Usually, however, a split plot which gives sufficient information on main plot treatments gives too much on sub-plots treatments, and it can only be justified because any alternative experiment would be larger.

Later we will see that there is a theoretical reason why comparisons of main plot treatments within one level of a sub-plot treatment are not at all straightforward. For the moment, note that a comparison of, say, windrowing and desiccating for direct drilled plots does not involve an estimate of main plot effects, whereas a comparison of direct drilling and roller drilling for windrowed plots does.

8.2.2 Some Theory

The new feature of the split plot model is that there are two linear models and therefore two variances. Treatment comparisons require different estimates of variances depending on whether they are within the same main plot (when the main plot effects cancel) or within different main plots (when the main plot and sub-plot effects both contribute) or between main plots (when the sub-plot effects will be averaged).

The full model is

$$y_{hij} = \mu + \beta_j + (\tau m)_i + \xi_{ij} + (\tau s)_h + (\tau ms)_{hi} + \varepsilon_{hij} \tag{8.2.3}$$

where $1 \leq h \leq l$, $1 \leq i \leq k$, $1 \leq j \leq r$

and $\Sigma\beta_j = \Sigma\tau m_i = \Sigma\tau s_h = \Sigma\tau ms_{hi} = 0$

and $\text{var}\ \xi = \sigma_m^2$, $\text{var}\ \varepsilon = \sigma_s^2$.

The variance of a main plot mean is

$$\text{var}\ \bar{y}_{.ij} = \text{var}\ \xi_{ij} + \text{var}\ \bar{\varepsilon}_{.ij}$$

$$= \sigma_m^2 + \sigma_s^2/l \tag{8.2.4}$$

The difference between two main plot treatment means therefore has variance

$$\begin{aligned} \text{var}(\bar{y}_{.r.} - \bar{y}_{.s.}) &= (\text{var } \bar{y}_{.rj})/r + (\text{var } \bar{y}_{.sj})/r \\ &= 2\sigma_m^2/r + 2\sigma_s^2/rl \end{aligned} \quad (8.2.5)$$

Because a main plot mean is the mean of l sub-plots, the main plot residual sum of squares, s_b^2 (b=between main plots), estimates

$$l\sigma_m^2 + \sigma_s^2 \text{ with } (r-1)\times(k-1) \text{ degrees of freedom.} \quad (8.2.6)$$

so that an estimate of the variance of the treatment difference (8.2.5) is $2s_b^2/(rl)$.

The variance of a sub-plot is

$$\begin{aligned} \text{var } y_{hij} &= \text{var } \xi_{ij} + \text{var } \varepsilon_{hij} \\ &= \sigma_m^2 + \sigma_s^2 \end{aligned} \quad (8.2.7)$$

The difference between two individual treatment means therefore has variance

$$\text{var}(\bar{y}_{ur.} - \bar{y}_{vs.}) = 2(\sigma_m^2 + \sigma_s^2)/r \quad (8.2.8)$$

unlesss r=s. If r=s, and the difference is between two sub-plot treatments at the same level of a main plot treatment, the ξ term cancels, and

$$\bar{y}_{ur.} - \bar{y}_{vr.} = (\tau s)_u - (\tau s)_v + \bar{\varepsilon}_{ui.} - \bar{\varepsilon}_{vi.}$$

and the variance is

$$\text{var}(\bar{y}_{ur.} - \bar{y}_{vr.}) = 2\sigma_s^2/r \quad (8.2.9)$$

Now σ_s^2 is estimated by s_w^2 (w = within main plots) with k(r-1)(l-1) degrees of freedom, so that an estimate of (8.2.9) is $2s_w^2/r$, but there is no direct estimate of (8.2.8). A weighted average can be formed,

$$((l - 1)s_w^2 + s_b^2)/(rl) \qquad \text{estimates} \quad \sigma_m^2 + \sigma_s^2$$

but this does not lead to exact inferences using a t distribution. A conservative procedure is to use a t-distribution with just the main plot degrees of freedom, $(r-1)(k-1)$, because this is always the less accurate of the two estimates. Finally, the variance of the difference between two sub-plot main effect means is

$$\text{var}(\bar{y}_{u..} - \bar{y}_{v..}) = \sigma_s^2/(rk) \qquad (8.2.10)$$

which is estimated by $2s_w^2/(rk)$.

Example 8.2.2 Calculation of standard errors

To complete Example 8.2.1 we will calculate variances of the various treatment differences. Tables of treatment means and their standard errors are given in Table 8.2.3. The standard error of a main plot mean is

$$\sqrt{(s_b^2/rl)} = \sqrt{(229017/12)} = 138$$

and this is the figure to use in any of the techniques of Chapter 6, Section 5 where main plot means are being compared. The standard error of a main plot treatment difference is

$$\sqrt{(2s_b^2/lr)} = \sqrt{(2\times 229017/12)} = 195$$

Direct drilling, with a mean yield of 1000 less than the other cultivation treatments, is clearly the inferior drilling method. One would not really expect the success of a harvesting method to depend on the way the crop was sown, and it did not. For each drilling method the pattern of differences between harvesting methods is much the same, and the interaction sum of squares was negligble. We can

TABLE 8.2.3 Means and Standard Errors for Cultivation Experiment

```
   ***** TABLES OF MEANS *****

            Cultn Aitchisn Dirct_dl   2_pass Roll_drl
                      2984     1919     2989     2981

          Harvest  Windrwd  Dessctd Drct_cut
                      3952     2647     1556

          Harvest  Windrwd  Dessctd Drct_cut
            Cultn
         Aitchisn     4437     2927     1589
         Dirct_dl     3213     1908      637
           2_pass     4135     2773     2059
         Roll_drl     4024     2980     1939

 ***** STANDARD ERRORS OF DIFFERENCES OF MEANS *****

 TABLE                   CULTN     HARVEST        CULTN
                                                HARVEST
 REP                        12          16            4
 SED                     195.4       270.1        482.4
 EXCEPT WHEN COMPARING MEANS WITH SAME LEVEL(S) OF:
   CULTN                                          540.2
```

therefore concentrate on the main effects. The standard error of a sub plot treatment mean is

$$\sqrt{(s_w^2/rk)} = \sqrt{(583595/16)} = 191,$$

and the standard error of a difference 270. Again the effects are clear cut, windrowed being better than desiccated, which in turn is better than untreated. If the interaction had been significant comparisons along the rows of the Harvest×Cultn table would be required, the standard error of a mean being $\sqrt{(583595/4)} = 382$, and of a difference 540. Comparisons between rows of this table involve the problems of using (8.2.10). Although the computer output shows this standard error as clearly as all the others, remember that it does not lead to exact t-tests.

8.3 REPEATED MEASURES

Often an experimental unit has the same type of observation made on it, but at different times. An observation may be made before treatment and then repeated after treatment. This may be the initial weight and final weight of an animal in an experiment comparing diets, or initial and final performance at a test in an experiment comparing methods of instruction. The animal scientist is usually quite content to analyse the weight gain as a single observation in a simple analysis, but the educationalist with social science training will prefer something more complicated. In agriculture, a series of harvests is sometimes made on the same plot, or a series of height measurements may be made on the plants in a plot.

In all these cases a split plot analysis is sometimes used or misused. The difficulty is that the repeated measure is not made on a separate experimental unit, and is not randomized. The split plot model might still be a reasonable description of the data, although without randomization the assumption that residuals are independent will be dubious. Also there is rather more likelihood of the variance changing as the observations will probably be made under rather different conditons. The criticism is rather that better procedures are usually available. These are

(i) Analyse a function of the repeated measures. For example, take the difference between a pre test and a post test. The mean and the difference of two independent observations span the same vector space as the observations themselves, and so are just a reparametrization of the same model. However the mean of a pre and post test measure has no practical interpretation, and so an analysis of the difference provides exactly the same practical inferences as a split plot.

Other examples might be the slope (or some other parameter) of a regression line for repeated observation in time, or the total, or even the maximum. There is no alternative but to decide what single measure is of importance, calculate it from the repeated measures, and analyse it as a single observation.

(ii) Do a separate analysis of each measure.

(iii) If the data is for successive observations in time, often the best way of presenting the results is as a graph against time. Each separate analysis gives the points on the graph, with a measure of their precision, for each time. If the graph is not sufficient, quite separate analyses can be performed to fit a model to changes in the observations through time.

(iii) If a grand analysis is required, it may be necessary to weight the observations differently to allow for the repeated measures having different variability. The weights would be the inverse of the residual variance from separate analyses. See Chapter 4, Section 4.

8.4 COVARIANCE ANALYSIS

Experimental units are inevitably variable. We have seen one way of coping with this problem in the chapter on blocking. Very often the variability of the experimental units can be measured. For example it is sometimes possible to make the same measurement before an experiment as it is intended to make after it, as with the initial and final weights of animals, or the pretest and post-test scores of social scientists' victims. It is reasonably likely that these two measurements will be correlated, so that including the initial measurement in the experimental design model will reduce the residual variability. Such a measurement is called a covariate. Any measurement made before the experimental units are assigned to treatments and which is correlated with the final observation can be tried as a covariate, and only experience can tell which covariates might give a useful reduction in the residual.

As a simple example of the calculations we will use a completely randomized design. The model is

$$y_{ij} = \mu + \tau_i + \varepsilon_{ij} \tag{8.4.1}$$

Adding a covariate, x, gives the model

$$y_{ij} = \mu + \tau_i' + \alpha x_{ij} + \varepsilon_{ij}' \tag{8.4.2}$$

The treatment parameters have been dashed (') because their values will change when the nonorthogonal **x** is added to the model. The normal equations could be solved in the usual way, but it is simpler to use the method of Chapter 3, Section 4. That is, fit **x** instead of **y** in (8.4.1), and then use the **x** residuals instead of **x** in (8.4.2). These residuals, labelled **z** to conform to Chapter 3, Section 4, are orthogonal to the treatment factors, so that model (8.4.2) becomes

$$y_{ij} = \mu + \tau_i + \alpha z_{ij} + \varepsilon'_{ij} \tag{8.4.3}$$

Orthogonality means that the estimates of the treatment effects are as in (8.4.1), and the estimate of α is the same as from a simple regression of **y** on **z**.

$$\begin{aligned}\hat{\alpha} = a &= \sum[(y_{ij} - \bar{y})\, z_{ij}] \,/\, [\, z_{ij}] \quad \text{since } \bar{z} = 0 \\ &= [\, \mathbf{y}^T \mathbf{z}\,]/[\, \mathbf{z}^T \mathbf{z}\,] \\ &= [(\, \mathbf{b}^T X^T + \mathbf{e}^T\,)\, \mathbf{z}]/[\, \mathbf{z}^T \mathbf{z}\,] \quad \text{since } \mathbf{y} = X\mathbf{b} + \mathbf{e} \\ &= [\, \mathbf{e}^T \mathbf{z}\,]/[\, \mathbf{z}^T \mathbf{z}\,] \quad \text{since } X^T \mathbf{z} = 0.\end{aligned}$$

The drop in residual sum of squares on fitting the covariate is

$$[\, \mathbf{e}^T \mathbf{z}\,]^2/[\, \mathbf{z}^T \mathbf{z}\,],$$

Therefore the residual from (8.4.2) can be found by subtracting this from the residual of (8.4.1).

This brief description shows that all the quantities required for a covariance analysis can be derived from an analysis of (8.4.1), first of the y's, then of the x's. Since the residuals are always orthogonal to the model, the same procedure will apply to any experimental design.

Example 8.4.1 Computer teaching data as covariance

The data Appendix C 6 could be analysed using the pretest figures as a covariate. The square root of the test scores has been used in this analysis to give a more uniform variability between groups. It was the improvement in word recognition which was important, and so the y variable was the difference between the two transformed scores. (The transformation means that an improvement from 2 to 4 gives y = 0.586, whereas an improvement from 6 to 8 gives y = 0.379. Whether this is reasonable is for the experimenter to judge.) There are three ANOVA tables one for the pretest (x) in Table 8.4.1, one for the difference (y) in Table 8.4.2 and one for y adjusted for x in Table 8.4.3. Points to note are

TABLE 8.4.1 Analysis of Pretest Scores (Squareroot Tranformation)

***** ANALYSIS OF VARIANCE *****

VARIATE: Pretest

SOURCE OF VARIATION	DF	SS	MS	VR
UNITS STRATUM				
Trtmnt	3	1.3782	0.4594	0.547
RESIDUAL	45	37.7934	0.8399	
TOTAL	48	39.1715	0.8161	

GRAND MEAN 1.08
TOTAL NUMBER OF OBSERVATIONS 49

***** TABLES OF MEANS *****

Trtmnt	A	B	C	D
	1.10	0.81	1.20	1.23
REP	12	13	12	12

TABLE 8.4.2 Analysis of Increase in Scores (Squareroot)

***** ANALYSIS OF VARIANCE *****

VARIATE: DIFFER

SOURCE OF VARIATION	DF	SS	MS	VR
UNITS STRATUM				
Trtmnt	3	2.641	0.880	0.805
RESIDUAL	45	49.246	1.094	
TOTAL	48	51.888	1.081	

GRAND MEAN 0.49
TOTAL NUMBER OF OBSERVATIONS 49

***** TABLES OF MEANS *****

Trtmnt	A	B	C	D
	0.26	0.39	0.45	0.89
REP	12	13	12	12

TABLE 8.4.3 Covariance Analysis of Increase in Scores (Squareroot)

```
***** ANALYSIS OF VARIANCE *****
(ADJUSTED FOR COVARIATE)

VARIATE: DIFFER

SOURCE OF VARIATION              DF         SS         MS        VR

*UNITS* STRATUM
  Trtmnt                          3     4.0660     1.3553     1.610
  COVARIATE                       1    12.2098    12.2098    14.505
  RESIDUAL                       44    37.0364     0.8417
TOTAL                            48    53.3122     1.1107

GRAND MEAN                           0.49
TOTAL NUMBER OF OBSERVATIONS           49

***** COVARIANCE REGRESSIONS *****

  COVARIATE                COEFFICIENT          SE
*UNITS* STRATUM
  Pretest                        -0.57       0.149

***** TABLES OF MEANS *****
(ADJUSTED FOR COVARIATE)

     GRAND MEAN     0.49
         Trtmnt        A         B         C         D
                    0.27      0.24      0.52      0.97
            REP       12        13        12        12
```

(i) The treatment sum of squares for PRETEST is small (VR = 0.547). Since the pretest was made before the random allocation of treatments, it would be suspicious if there were any sign of treatment effect.

(ii) Including a covariate has lowered the residual mean square, from 1.094 to 0.842.

(iii) The treatment mean square has changed only slightly on including the covariate. We would not expect much change because although the covariate is not orthogonal to treatments, it should be stochastically independent of them.

(iv) The covariate sum of squares is the increase in residual sums of squares resulting from omitting the covariate from the full model (8.4.3), and likewise the treatment sum of squares is the increase on omitting the treatment factor. These two do not add to anything in particular, and so the total sum of squares in the adjusted ANOVA does not equal the total sum of squares for DIFFER.

8.4.1 The Adjusted Treatment Means

The random application of treatments will result in the treatment groups from the experiment having different means for the covariate.

The aim of covariance is to remove the effect of these differences, and equation (8.4.2) can be used to predict what the treatment means would have been if each treatment had had the same value of the covariate. An obvious choice of covariate value to predict at is the mean value, $\bar{x}_{..}$. The corresponding value for z for the i-th treatment is $\bar{x}_{..}-\hat{x}_{ij}$, which for the completely randomized design is $\bar{x}_{..}-\bar{x}_{i.}$. From (8.4.3) then the predicted mean of the i-th treatment is

$$\mu + \tau_i + \alpha(\bar{x}_{ij} - \bar{x}_{i.})$$

which is estimated by

$$\bar{y}_{i.} + a(\bar{x}_{..} - \bar{x}_{i.}) \qquad (8.4.4)$$

The adjusted estimate of the difference between two treatments is

$$(\bar{y}_{i.} - \bar{y}_{j.}) - a(\bar{x}_{i.} - \bar{x}_{j.})$$

Example 8.4.2 Adjusted treatment estimates

Tables of means were included in the Tables 8.4.1, 2 and 3. Substituting these means into (8.4.4) for treatment A gives

$$\text{Adjusted mean for A} = 0.26 - (-0.57)\times(1.10-1.08)$$
$$= 0.27$$

which agrees with the mean in Table 8.4.3.

Note that:

(i) The coefficient a is negative, so that those scoring low at first tended to improve more.
(ii) The effect of the adjustment is to decrease those who scored low at first (e.g. group 4), to counter the effect mentioned in (i).

The variance of these adjusted means is found from (8.4.4) to be

$$\text{var}(m+t_i) = \text{var } \bar{y}_i \text{ (adj)} = \text{var } \bar{y}_i + (\bar{x}_i - \bar{x})^2 \text{var } a$$

Therefore

$$\text{var}(\bar{y}_i - \bar{y}_j)(\text{adj}) = \text{var } \bar{y}_i + \text{var } \bar{y}_j + (\bar{x}_i - \bar{x}_j)^2 \text{ var } a$$

Each pair of means has a different variance, which makes inferences tedious, but both var a and the differences between the x means are usually small. Inferences using the largest and smallest differences between the x means will therefore be much the same.

As a means of increasing precision, covariance is very like blocking. Covariates, like blocks, are determined before the random allocation of treatments, ensuring that the covariate will be independent of treatment effects. Also like blocks, any interactions between covariate and treatment will be allocated to residual variability. Sometimes however a variable measured during the experiment is used as a covariate. For example the treatment might be the rate of flow of liquid through a reaction, and the temperature of the reaction might be the "covariate". The observation might be the yield of desired product at the end of the reaction. Possible paths for the variables to affect each other are shown in Figure 8.4.1. If the temperature affects the yield linearly (path (c)) and the flow rate affects the yield only by changing the temperature (i.e. path (a) is negligible), then a covariance analysis will show no treatment effect, because it is absorbed into the covariate. There will however be a flowrate effect on temperature, something which would not happen in a standard covariance analysis because in it the covariate is measured before treatments are applied, so that the way the experiment is carried out ensures that path (b) must be zero. It may be very useful to find out the extent to which flow rate affects yield directly, but the assumption that the relationship (c) be linear is crucial. If there were an optimum temperature, and yield fell away at both high and low temperatures, the temperature might well

```
              ------->   Temperature   ------->
Flow rate        b       (covariate)      c          Yield
(treatment)   ---------------------------------->  (observation)
                              a
```

FIGURE 8.4.1 Possible paths for flow rate to affect yield.

not be significant even though the cause and effect was working entirely along paths (b) and (c). Unfortunately, if flowrate affected both temperature and yield, but temperature itself had little effect on yield (path(c) negligible) a covariance analysis would give the same results as if path(a) were small and paths (b) and (c) large. There is no way any analysis can sort out the causal relationships between variables.

Sometimes covariance is used as a means of estimating treatment differences when treatments are not allocated randomly. The method is to measure a few key variables on each experimental unit and use these as covariates. The hope is that the effects of all differences between treatment groups other than treatments themselves are removed by the covariates. The assumptions are that the key variables measured summarize all the factors which might otherwise have been confounded with treatments, and that any effects they have are linearly related to the observations.

8.5 UNEQUAL REPLICATION

We have already noted that unequal replication means that factorial effects, interactions, and orthogonal contrasts are not independent. Orthogonality of effects no longer implies orthogonality of the columns of the X matrix. As we know from the regression theory of Chapter 3, the estimates for one term in the model will depend on which other terms are in the model. We can no longer talk of *the* sum of squares for an effect. The analysis becomes rather that of Chapter 3, although because the data comes from a designed experiment the full model is known. Also the fundamental requirements of randomization, replication, and control of variability are in no way affected. We will demonstrate the problem with an example.

Example 8.5.1 Unbalanced experiment

We will now analyse the computer teaching experiment (Appendix C 6) including the order factor. The experimenter was not expecting this factor to have any effect, but we will include it in this analysis. Again in this analysis the pretest and post-test scores will be given

TABLE 8.5.1 Analysis as Eight Separate Treatments

```
***** ANALYSIS OF VARIANCE *****

VARIATE: DIFFER

SOURCE OF VARIATION              DF        SS        MS       VR

*UNITS* STRATUM
  Trtmnt                          7      7.899     1.128     1.05
  RESIDUAL                       41     43.989     1.073
TOTAL                            48     51.888     1.081
GRAND TOTAL                      48     51.888
GRAND MEAN                          0.49
TOTAL NUMBER OF OBSERVATIONS          49

***** TABLES OF MEANS *****

    GRAND MEAN      0.49

     Trtmnt      1      m      t     mt      o     mo     to    mto
              0.81   0.52   0.68   1.00  -0.29   0.29  -0.02   0.77
        REP      6      6      8      6      6      7      4      6

***** STANDARD ERRORS OF DIFFERENCES OF MEANS *****

TABLE                Trtmnt
---------------------------
REP                  UNEQUAL
SED                    0.732X MIN REP
                       0.634  MAX-MIN
                       0.518X MAX REP
```

a square root transformation. First the experiment is treated as one with eight individual treatments. The results are given in Table 8.5.1. From these treatment means contrasts can be calculated for the main effects and interactions, and the EMS can be used in tests of significance or confidence intervals. There are no differences in principle from the procedures described in Chapter 6, although the r's can no longer be taken out as a common factor making the computations rather more tedious. For the order effect we have

$$O = [(0.81 + \ldots + 1.00) - (-0.29 + \ldots + 0.77)]/8$$
$$= 0.283$$
$$\text{var } O = \{[(1/6 + \ldots + 1/6) + (1/6 + \ldots + 1/6)]/64\} \times 1.073$$
$$= 0.02265$$

so that

$$t = 0.283/\sqrt{0.02265}$$
$$= 1.88$$

which might be large enough to cause the experimenter to have second thoughts about the order effect.

To analyse the data as a 2×2×2 factorial with an ANOVA table displaying sums of squares for each factor, we transform the design model from one in terms of τ to one in terms of γ by following (6.3.7) and (6.3.8). The resulting X matrix is shown in Table 8.5.2. The columns of the matrix are not orthogonal. The consequences of this are shown in the set of computer output in Table 8.5.3. Our analysis has been performed using a regression program, not an experimental design program. The output gives the estimates of the three main effects and the seven interaction parameters with their standard errors.

Where seven contrasts are fitted corresponding to the full set of main effects and interactions (Model I), the estimates and their standard errors are the same as those calculated from the individual treatments. The second table is for a model omitting the three factor interaction (Model II), and the third is a model with no interactions at all (Model III). As predicted, the estimates do change, although not very greatly because the sample sizes are not very

TABLE 8.5.2 X Matrix for Teaching Experiment Factors

Effects								
Methd	Teach	Order	MxT	TxO	MxO	MxTxO	Trtmnt	r_i
+1	−1	−1	+1	+1	+1	−1	(1)	8
+1	−1	−1	−1	+1	−1	+1	t	6
−1	+1	−1	−1	−1	+1	−1	m	6
+1	+1	−1	+1	−1	−1	+1	tm	6
−1	−1	+1	+1	−1	−1	−1	o	4
+1	−1	+1	−1	−1	+1	+1	to	6
−1	+1	+1	−1	+1	−1	−1	mo	6
+1	+1	+1	+1	+1	+1	+1	tmo	7

TABLE 8.5.3 Regression Analysis of Teaching Data

***** REGRESSION ANALYSIS ***** Model I - Full Model
Y-VARIATE: DIFFER
*** REGRESSION COEFFICIENTS ***

	ESTIMATE	S.E.	T
CONSTANT	0.470	0.151	3.12
Method	0.174	0.151	1.15
Teacher	0.138	0.151	0.92
Order	-0.283	0.151	-1.88
TxO	0.049	0.151	0.32
OxM	0.167	0.151	1.11
MxT	0.103	0.151	0.69
OxTxM	-0.048	0.151	-0.32

***** REGRESSION ANALYSIS ***** Model II - omitting OxTxM
Y-VARIATE: DIFFER
*** REGRESSION COEFFICIENTS ***

	ESTIMATE	S.E.	T
CONSTANT	0.466	0.148	3.14
Method	0.179	0.148	1.21
Teacher	0.133	0.148	0.90
Order	-0.284	0.149	-1.91
TxO	0.046	0.149	0.31
OxM	0.169	0.149	1.14
MxT	0.107	0.148	0.72

***** REGRESSION ANALYSIS ***** Model III - main effects only
Y-VARIATE: DIFFER
*** REGRESSION COEFFICIENTS ***

	ESTIMATE	S.E.	T
CONSTANT	0.478	0.145	3.30
Method	0.168	0.145	1.16
Teacher	0.144	0.145	0.99
Order	-0.272	0.146	-1.86

different. However in other cases the changes could be substantial. The problem is, which estimates should be used and which models should be used for hypothesis tests? There is no problem with the three factor interaction as it can only be estimated by comparing Model I with Model II. But there are three possible t-values in Table 8.5.3 for testing the order effect, and others could be constructed. Different computer programs and different statisticians would use different procedures. The two conflicting principles are:

(i) If any terms are wrongly omitted from the model, the parameter estimates are biased by an amount depending on the size of the omitted parameters. If Model III were used to test the hypothesis that the order effect is zero, what is in fact being tested is that the sum of the order effect and some combination of the interactions involving order is zero. The specific details can be found in Speed and Hocking, 1976. We know that Model I is the full model, so that if it is used the estimates will be unbiased.

(ii) If a main effect is not in the model, interactions involving it do not make much sense, as was explained in Chapter 6, Section 4. If Model I is used for testing the order effect, a model with the order effect is being compared with a model which excludes it but includes all its interactions. Model III contains only main effects, so that this problem does not arise using it.

The points at issue are too fundamental to be resolved here, but at a practical level the problem can be partially resolved by recalling that a hypothesis test gives positive evidence for the presence of a factor, but a nonsignificant parameter is not necessarily a zero parameter. We suggest that it is sensible to use a two stage procedure.

(i) Estimation. Before the experiment the teacher and method main effects and their interaction were thought to be important. The experiment itself suggested that the order effect could not be ignored. None of the interactions with order were expected to be important, and the experiment reinforces this expectation. We could leave these out and use a model with the three main effects and the teacher by method interaction with little chance of the estimates being biased.

For estimation then we require evidence that an effect is zero before leaving it out, and we would certainly not include a term without including all terms marginal to it. This principle was used in Chapter 7, Section 4.1 when the high order interactions were used to provide an estimate of σ^2.

(ii) Hypothesis tests. Having settled on a model which includes all parameters likely to be nonzero, the next question is, which parameters does this experiment give positive evidence for being nonzero. If the model constructed in (i) is used as the basis for these tests, main effects will be tested by comparing this model with one excluding a main effect but including interactions involving it. As this is a process for assessing the evidence for the main effect, its use does not imply that the main effect is in fact zero.

In Chapter 3, Section 7, much more stringent requirements were used to determine whether a term should be in the model, but there the number of independent variables was unbounded. This is quite unlike an experiment, where the number of independent variables is limited by the number of treatments, and each one is known to represent differences in the way the experimental units were treated.

8.5.1 Missing Data

Often a balanced experiment will suffer an accident which makes it unbalanced. Animals may die, technicians may make mistakes or the weather may change. If only a few experimental units are affected it is easier to carry out a special missing data analysis than to treat the experiment as unbalanced. This involves an iterative technique to estimate values to replace those missing by minimising the residual sum of squares. An initial estimate, perhaps the block mean, is substituted for each missing value, and the standard analysis is performed. Then the initial estimate is replaced with the predicted value and the procedure is repeated until there is no further reduction in the residual sum of squares. One residual degree of freedom must be subtracted for each missing plot. Since the residual sum of squares is minimised the estimates of residual variance and the parameters will be the same as the least squares estimates from the unbalanced data, but the sums of squares for individual effects will not be be quite the same, and only approximate estimates will be given for standard deviations of parameter estimates.

8.6 MODELLING THE DATA

So far, in our accounts of experimental design we have assumed that all treatments have been properly randomized. Randomization should never be abandoned lightly, but it can happen that a properly randomized design is impossibly extravagant of space or time. Or, more commonly, the experimenter believes it to be impossible and starts before a suitably persuasive statistician arrives. In other cases a properly randomized experiment might turn out to have a pattern of treatments which clearly favours one treatment over another. As we said in Chapter 5 any effect of this kind would average to zero over a series of experiments, and the probability of it leading to a significant result is accurately measured by the probability level of the test. This is small comfort when there has been a struggle to raise the resources for just one experiment.

In both these cases it may be necessary to abandon the strict ideal of performing the analysis required by the design and instead choose an analysis which allows for the peculiarities of the data. A distinction can be made between _confirmatory_ analysis, which answers (hopefully confirms) the suspicions the experiment was designed to investigate, and _exploratory_ analysis, which digs around in the data looking for information which was not initially suspected. The procedures described in Chapters 3 and 4 when applied to experimental data are really exploratory analyses.

Example 8.6.1 Non-randomized potato experiment

In 1930 a New Zealand crop experimentalist, E. A. Hudson, had problems with the analysis of a potato experiment. He was comparing certified seed with noncertified seed in an experiment which comprised 20 plots with the different types of seed in alternate plots, not randomized. The order is shown in the analysis which follows. When he carried out a paired t-test with plots 1v2, 3v4, ..., 19v20 the difference was significant, but when he paired them 2v3, 4v5,..., 18v19 the difference was not significant. (He also obtained a highly significant result by pairing the 19 differences 1v2, 2v3,..., 19v20 as independent pairs, but we need not comment on that.) Hudson wrote to W. S. Gosset ("Student") for help. Gosset did not refuse to have anything to do with this poorly designed experiment. He was not a great believer in randomization, and considered that an experienced experimenter could choose an arrangment in such a way that other effects (soil fertility trends, for example) cancelled. Clearly Hudson's experiment did not meet this criterion as plots yields decreased from left to right. Each right hand plot was therefore a little better than the one on its left, which added to the treatment effect when the pairing was 1v2, 2v3, 19v20, but subtracted from it when the pairing was 2v3, 4v5, 18v19. In fact when blocked in pairs the treatment effect is confounded with a linear fertility trend. The analysis carried out by Gosset's assistant, E. Sommerville, is reproduced below. In essence he assumed that there was a linear fertility which he estimated separately for each treatment using linear regression. Then he tested whether the difference between the NC and C plots was greater than would be predicted by this trend. The result was inconclusive.

> There is an obvious fertility slope in this experiment and with such an arrangement of plots the results must give an exaggerated idea of the real difference. We can eliminate the effect of this fertility slope if we fit a straight line to the plots (actually a straight line was fitted to each of plots - N and NC - and the average slope taken) and analyse the variance or the deviations of plots from the straight line value, as follows:-

	Actual values	Straight line values	Difference	
C	17.5	19.0	-1.5	} -2.3
NC	18.0	18.8	- .8	
C	18.7	18.5	.2	} -2.5
NC	15.6	18.3	-2.7	
C	22.1	18.0	4.1	} 3.9
NC	16.7	17.8	-1.1	
C	22.6	17.5	5.1	} 3.2
NC	15.4	17.3	-1.9	
C	14.5	17.0	-2.5	} -4.9
NC	14.4	16.8	-2.4	
C	17.3	16.5	.8	} 5.1
NC	20.6	16.3	4.3	
C	15.6	16.0	- .4	} - .4
NC	15.8	15.8	.0	
C	17.4	15.5	1.9	} 2.8
NC	16.1	15.2	.9	
C	14.1	15.0	-0.9	} -4.0
NC	11.6	14.7	-3.1	
C	15.6	14.5	1.1	} .0
NC	13.1	14.2	-1.1	
	332.7	332.7	0	

Mean difference C and NC = 1.58

Variance due to:-	Degrees of Freedom	Sum of Squares	Mean Square	Standard Deviation
Blocks	8	52.40		
Treatment	1	12.48		
Random	9	41.54	4.62	2.15
Total	18	106.42		

$$t = \frac{1.58 \times \sqrt{10}}{\sqrt{2} \times 2.15} = 1.65 \qquad P = .933$$

giving odds of 14 to 1

(E.Somerville)

Gosset's analysis differed from our usual analysis because he had to make an assumption about the fertility effect before he could estimate the treatment effect. If the fertility effect was not in fact linear, his estimate would be biased - see Chapter 3, Section 6. The effect of randomization is to turn all factors not allowed for in the design into random effects which therefore cause no bias. We may use covariance or transformations to reduce the residual variance or to make the residuals fit some standard pattern, but the treatment effects are always estimated unbiasedly without them. Note that both Hudson and Gosset laid down many field experiments, and a bias introduced by a wrong assumption about a fertility trend would mean that there could be a consistent error in all their estimates.

For ten years Hudson spurred Gosset on in the battle against randomization. Finally Gosset wrote a paper ("Student", 1937) which showed that randomized designs could lead to less precise estimates than systematic designs. Yet in spite of this the randomizers led by Fisher and Yates won the debate. But Hudson and Gosset had an important lesson to teach. A close knowledge of ones experimental material, carefully built into the design of an experiment using the techniques of this book will provide the maximum amount of information with the least effort. Further, the existence of the computer permits a close examination of experimental results in a way which neither Gosset nor Fisher would have dreamt of. Such an exploration provides a wide variety of leads for future experiments, but the information so obtained cannot be given much weight as information about the population parameters.

PROBLEMS

8.1. In an industrial extrusion process there are three factors affecting the strength of the product: temperature, speed, and pressure. To make an initial study of the effect of these factors a 2×2×2 experiment was carried out as a split plot with temperature as main plot treatment. The strength of the product resulting from each run is given in the table below.

Day 1, morning:	Treatment	Tps	TPS	TpS	TPs
	Strength	110	104	112	102
afternoon:	Treatment	tps	tPs	tpS	tPS
	Strength	112	110	107	112
Day 2, morning:	Treatment	tPS	tPs	tpS	tps
	Strength	115	109	113	110
afternoon:	Treatment	TPs	TpS	Tps	TPS
	Strength	103	116	111	107

Analyse the results to determine which effects are significant.

8.2. Assumptions required for a splitplot experiment are that the ξ_i's are all independent and that the ε_i's are independent within each main plot. Discuss how well these assumptions are likely to be satisfied in the following cases.

(i) A field experiment laid out as below with sub-plots vertical strips.

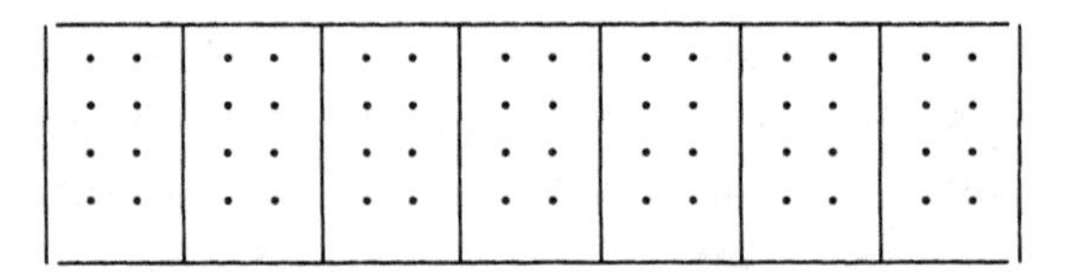

(ii) A field experiment laid out as below with sub-plots square

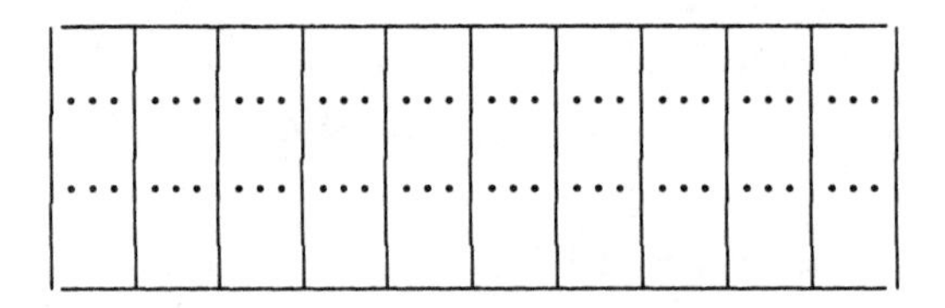

(iii) An industrial experiment where operator skill is an important factor. Main plot treatments were randomized over days, and sub-plot treatments were randomized within days. There were two operators, one for runs in the morning, the other for the afternoons.

(iv) Two factors which could be important in the quality of teaching a skill are the teacher's understanding of the teaching problems, and the technique used to present it. A selection of different explanations of the teaching problems are randomly distributed to individual teachers (main plots). Then each teacher's class is randomly divided into groups, which become sub-plots. Teachers use a different technique for each group in their classes. Finally some numerical assessment is made of the level of skill of each group of children.

8.3. One of the several experiments comparing yields resulting from various harvesting and cultivation techniques was wrongly analysed as a randomized complete block. In fact the harvesting treatments were sub-plot treatments. The incorrect analysis was as given below. Carry out the correct analysis. Describe the changes in interpretation resulting from the new analysis. Are they what might have been expected?

HARVEST	CULTN REPS	1	2	3	4	5	6	TOTAL
1	1	1846	1043	1629	1483	1688	1796	9485
	2	1875	1489	1756	1182	1218	2137	9657
	3	2337	1590	2242	2213	1943	1244	11569
	4	1977	1577	1496	2094	1616	1726	10486
	TOTAL	8035	5699	7123	6972	6465	6903	41197
2	1	1319	646	1526	1033	1006	1112	6642
	2	1027	942	1177	812	1078	1415	6451
	3	1136	828	1436	1939	1152	1034	7525
	4	1836	1114	1030	1609	1079	1187	7855
	TOTAL	5318	3530	5169	5393	4315	4748	28473
3	1	946	831	983	977	153	190	4080
	2	444	98	788	1175	163	265	2933
	3	782	36	764	534	456	154	2726
	4	499	105	826	660	257	203	2550
	TOTAL	2671	1070	3361	3346	1029	812	12289
TOTAL	1	4111	2520	4138	3493	2847	3098	20207
	2	3346	2529	3721	3169	2459	3817	19041
	3	4255	2454	4442	4686	3551	2432	21820
	4	4312	2796	3352	4363	2952	3116	20891
	TOTAL	16024	10299	15653	15711	11809	12463	81959

***** ANALYSIS OF VARIANCE *****

VARIATE: OIL_YLD

SOURCE OF VARIATION	DF	SS	MS	VR
REPS STRATUM	3	228300	76100	
REPS.*UNITS* STRATUM				
CULTN	5	2493540	498708	5.850
HARVEST	2	17492980	8746490	102.607
CULTN.HARVEST	10	725583	72558	0.851
RESIDUAL	51	4347387	85243	
TOTAL	68	25059492	368522	
GRAND TOTAL	71	25287792		

8.4. In the teaching experiment, Appendix C 6, either the post-test scores or the improvement could have been analysed. Show that if the pretest score is used as a covariate these two analyses will lead to the same residuals.

8.5. Cream that has been frozen is not suitable for whipping without the addition of emulsifiers and stabilizers. One measure of the quality of whipped cream is the amount of liquid which drains off it when a sample is placed in on filter paper for 24 hours. This is termed the leakage, and the smaller it is the better. The data below comes from an experiment to study the effect of two factors, homogenization pressure (6.9 MPa and 13.8MPa) and the stabilizer genulacta (none, 0.05%, 0.1%). The intention was to manufacture three samples with each treatment combination, but time ran out before the last two samples could be completed. Two analyses are shown. One uses a missing value analysis as described in Section 5.1 and the other shows the results of fitting regression models as in Section 5. Compare these two analyses, particularly the sums of squares they give for each factor and for each effect. Summarize the conclusions to be drawn from the experiment.

Factors			Contrasts				
PRESSURE	GENLCTA	LEAKAGE	L_PRESS	L_GEN	Q_GEN	L_GNxPR	Q_GNxPR
6.9	NONE	0.3	-1	-1	1	1	-1
6.9	NONE	0.4	-1	-1	1	1	-1
6.9	NONE	0.5	-1	-1	1	1	-1
6.9	0.05	*	-1	0	-2	0	2
6.9	0.05	0.7	-1	0	-2	0	2
6.9	0.05	0.9	-1	0	-2	0	2
6.9	0.10	0.4	-1	1	1	-1	-1
6.9	0.10	0.3	-1	1	1	-1	-1
6.9	0.10	0.3	-1	1	1	-1	-1
13.8	NONE	1.3	1	-1	1	-1	1
13.8	NONE	1.6	1	-1	1	-1	1
13.8	NONE	1.1	1	-1	1	-1	1
13.8	0.05	0.9	1	0	-2	0	-2
13.8	0.05	0.4	1	0	-2	0	-2
13.8	0.05	0.8	1	0	-2	0	-2
13.8	0.10	0.8	1	1	1	1	1
13.8	0.10	0.5	1	1	1	1	1
13.8	0.10	*	1	1	1	1	1

***** REGRESSION ANALYSIS *****

*** REGRESSION COEFFICIENTS ***

	ESTIMATE	S.E.	T
CONSTANT	0.7028	0.0482	14.58
L_PRESS	0.1917	0.0482	3.98
L_GEN	-0.1875	0.0580	-3.23
Q_GEN	-0.0236	0.0347	-0.68
L_GNxPR	-0.1542	0.0580	-2.66
Q_GNxPR	0.1208	0.07	3.48

*** ANALYSIS OF VARIANCE ***

	DF	SS	MS
REGRESSN	5	1.9017	0.38033
RESIDUAL	10	0.3583	0.03583
TOTAL	15	2.2600	0.15067

```
***** ANALYSIS OF VARIANCE *****   (Using missing value analysis)
VARIATE: LEAKAGE
SOURCE OF VARIATION            DF(MV)         SS        MS         VR
*UNITS* STRATUM
  PRESSURE                       1         0.66125   0.66125    18.453
    LIN                          1         0.66125   0.66125    18.453
  GENLCTA                        2         0.44194   0.22097     6.167
    LIN                          1         0.42187   0.42187    11.773
    QUAD                         1         0.02007   0.02007     0.560
  PRESSURE.GENLCTA               2         0.81083   0.40542    11.314
    LIN.LIN                      1         0.28521   0.28521     7.959
    LIN.QUAD                     1         0.52562   0.52562    14.669
  RESIDUAL                      10( 2)     0.35833   0.03583
TOTAL                           15         2.27236   0.15149

ESTIMATED GRAND MEAN                   0.703
TOTAL NUMBER OF OBSERVATIONS              18
NUMBER OF MISSING VALUES                   2
MAXIMUM NUMBER OF ITERATIONS               2

          UNIT      ESTIMATED
        NUMBER          VALUE
             4          0.800
            18          0.650
```

```
***** TABLES OF EFFECTS *****
*** *UNITS* STRATUM ***
  PRESSURE CONTRASTS:
    LIN                        ESTIMATE    0.192     SE   0.0446
  GENLCTA CONTRASTS:
    LIN                        ESTIMATE   -0.187     SE   0.0546
    QUAD                       ESTIMATE   -0.0240    SE   0.0315
  PRESSURE.GENLCTA CONTRASTS:
    LIN.LIN                    ESTIMATE   -0.154     SE    0.0546
    LIN.QUAD                   ESTIMATE    0.121     SE    0.0315
```

```
***** TABLES OF MEANS *****
      PRESSURE      6.9      13.8
                  0.511     0.894

       GENLCTA      0.0       .05       .10
                  0.867     0.750     0.492

       GENLCTA      0.0       .05       .10
      PRESSURE
           6.9    0.400     0.800     0.333
          13.8    1.333     0.700     0.650
```

APPENDIX A
REVIEW OF VECTORS AND MATRICES

The following review summarizes common properties of vectors and matrices often used in statistics and are merely collected together here for convenience. Most of these properties are referred to in the body of the text and we indicate where each property was first employed.

A.1 SOME PROPERTIES OF VECTORS

1. In n dimensions, a vector **a** consists of n components of a_1 through a_n. Vector **a** is written in bold type. We shall usually consider **a** to be a column vector.

$$\mathbf{a} = \begin{vmatrix} a_1 \\ a_2 \\ \vdots \\ a_n \end{vmatrix}$$

The transpose of **a** is a row vector written as

$$\mathbf{a}^T = (a_1, a_2, \ldots, a_n)$$

Both **a** and $\mathbf{a}^T$ can be thought of as representing a point in n dimensional space, or as representing the direction from the origin to **a**. In comparison, the usual numbers we employ, and which form the elements of vectors, are called scalars. To distinguish between vectors and scalars, we shall denote vectors by bold type.

We often use **x** and **y** to denote the vectors of the predictor and dependent variables, respectively, as in Chapter 1, Section 5.

2. Addition of vectors is performed componentwise. If $\mathbf{c} = \mathbf{a} + \mathbf{b}$, then

$$\begin{vmatrix} c_1 \\ c_2 \\ \cdot \\ c_n \end{vmatrix} = \begin{vmatrix} a_1 \\ a_2 \\ \cdot \\ a_n \end{vmatrix} + \begin{vmatrix} b_1 \\ b_2 \\ \cdot \\ b_n \end{vmatrix} = \begin{vmatrix} a_1 + b_1 \\ a_2 + b_2 \\ \cdot \\ a_n + b_n \end{vmatrix}$$

Multiplication of a vector by a scalar is again performed componentwise so that (1.5.3) could be written as

$$\mathbf{x}^T b = (x_1, x_2, \ldots, x_n)\, b = (bx_1, bx_2, \ldots, bx_n)$$

3. The inner, or scalar, product of two vectors **a** and **b** is $\mathbf{a} \cdot \mathbf{b}$ or

$$\mathbf{a}^T \mathbf{b} = (a_1 b_1 + a_1 b_2 + \cdots + a_n b_n)$$

This product is also performed componentwise.

4. The length of a vector, $||\mathbf{a}||$, is defined as

$$||\mathbf{a}||^2 = (a_1^2 + a_2^2 + \cdots + a_n^2) = \mathbf{a}^T \mathbf{a}$$

We are often more concerned with the square of the length than the length itself, as this represents the sum of squares, or sum of cross products, as in the normal equations of (1.5.8) or the ANOVA tables of Chapter 2, Section 4.

5. $k\mathbf{a}$, where k is a constant, is a vector in the direction of **a** with a length $|k| \times ||\mathbf{a}||$.

Notice that $\mathbf{a}/\sqrt{[\mathbf{a}\cdot\mathbf{a}]}$ is a vector in the direction of **a** of length equal to one, that is, it is a unit vector in the direction of **a**.

6. If θ is the angle between two vectors **a** and **b** then the cosine of θ is given by

$$\cos\theta = (\mathbf{a}\cdot\mathbf{b})/\sqrt{[(\mathbf{a}\cdot\mathbf{a})(\mathbf{b}\cdot\mathbf{b})]}$$

Notice that the cosine could be written as the scalar product of two unit vectors, one in the direction of **a** and the other in the direction of **b**. If $\cos\theta = 0$, the vectors are said to be orthogonal (or at right angles).

The quantity $\cos\theta$ is very important in statistics. When the variables are expressed as deviations from their means it is the correlation coefficient between the two vectors. This is introduced in Chapter 1, Section 6.2.

7. A vector **c** can be <u>uniquely</u> written as the sum of two component vectors **a** and **b** when only the directions of **a** and **b** are specified. This property complements Property 2 above. In particular, we are usually concerned with splitting a vector into orthogonal parts. Thus, **c** can be uniquely decomposed into a vector in the direction of **b** and another in a direction orthogonal to **b**. We are often interested in splitting the dependent vector **y** into $\hat{\mathbf{y}}$, the predicted vector, and **e**, the vector of residuals as in Chapter 1, Section 5.

8. The vector in the direction of **b** can be shown to have the value $P\mathbf{c}$ where P, or P(**b**), is called the projection matrix onto **b** and (I-P) is the projection matrix in a direction orthogonal to **b**. It can be shown that

$$P = \mathbf{b}\,[\mathbf{b}^T\mathbf{b}]^{-1}\mathbf{b}^T = \mathbf{i}\,\mathbf{i}^T$$

where **i** is a unit vector in the direction of **b**. Clearly,

$$P^2 = PP = \mathbf{i}\,\mathbf{i}^T\,\mathbf{i}\,\mathbf{i}^T = P$$

as $\mathbf{i}^T\,\mathbf{i} = 1$ and $P(I-P) = P - PP = P - P = 0$

Using the properties of the transpose of a matrix, it is clear that P is symmetric and the two vectors $P\mathbf{c}$ and $(I-P)\,\mathbf{c}$ are orthogonal as required as

$$(P\mathbf{c})^T(I-P)\mathbf{c} = \mathbf{c}^T P^T (I-P)\,\mathbf{c} = 0$$

When the method of least squares is used, regression can be thought of as projecting the dependent variable onto the line, or plane, defined by the predictor variables.

A.2 SOME PROPERTIES OF VECTOR SPACES

In this text we do not formally discuss very much about vector spaces, but they are inherent in such discussions as in Chapter 3, Section 3 where it is pointed out that any two vectors which are not in the same direction define a plane. More formally, we could say that any two noncollinear vectors form a basis for a vector space of rank two.

We now define these terms more carefully. Unless stated otherwise, each vector is a column vector of n elements.

1. $\mathbf{b}$ is (linearly) dependent on $\mathbf{a}_1, \mathbf{a}_2, \ldots, \mathbf{a}_m$ and lies in the space defined by these vectors if there are constants c_i such that

$$\mathbf{b} = c_1\mathbf{a}_1 + c_2\mathbf{a}_2 + \cdots + c_m\mathbf{a}_m$$

Conversely, $\mathbf{b}$ is (linearly) independent of the $\mathbf{a}$ vectors if no such constants can be found.

In particular, we are concerned that the predictor variables, the $\mathbf{x}_i$, are linearly independent. If not, the problem of multicollinearity arises as explained in Chapter 4, Section 6. Also, $\hat{\mathbf{y}}$ is linearly dependent on the $\mathbf{x}$ variables.

2. The set of vectors dependent on any set of vectors, the $\mathbf{a}$'s, say, is called the vector space <u>generated</u> by the $\mathbf{a}$'s. If the generating vectors form an independent set, they are said to form a <u>basis</u> of the vector space.

3. The rank of a vector space is r if the number of vectors forming a basis of the space is r. A basis will not be unique as any number of sets of r independent vectors could be chosen as a basis. If the rank is r, however, each basis will consist of r independent vectors.

4. If the vector $\mathbf{b}$ is orthogonal to each of the $\mathbf{a}$'s, (that is $\mathbf{b}^T\mathbf{a}_i = 0$ for $i = 1,2,\ldots, m$) then $\mathbf{b}$ is orthogonal to the vector space generated by these vectors.

5. If the $\mathbf{a}$'s form a basis for a vector space, then any vector $\mathbf{b}_j$ in the space can be uniquely written as

$$\mathbf{b}_j = \sum c_{ij}\mathbf{a}_i \quad \text{where the } c_{ij} \text{ are constants}$$
$$= A\,\mathbf{c}_j \quad \text{where}$$
$$\mathbf{c}_j^T = (c_{1j}, c_{2j}, \ldots, c_{mj}) \text{ and the matrix}$$
$$A = (\mathbf{a}_1, \mathbf{a}_2, \ldots, \mathbf{a}_m)$$

6. If the $\mathbf{a}$'s form one basis for a vector space then another basis of $\mathbf{b}$'s, say, could be formed such that the $\mathbf{b}$'s form an orthogonal set, that is

$$\mathbf{b}_i^T\,\mathbf{b}_j = 0 \quad \text{if} \quad i \neq j$$

We can write this statement in matrices if we let $A = (\mathbf{a}_1, \ldots, \mathbf{a}_m)$ and $B = (\mathbf{b}_1, \ldots, \mathbf{b}_m)$ and the matrix of constants, C, be

$$C = \begin{vmatrix} c_{11} & c_{12} & \cdots & c_{1m} \\ \vdots & \vdots & & \vdots \\ c_{m1} & c_{m2} & \cdots & c_{mm} \end{vmatrix}$$

Thus $B = AC$ and $B^TB = D$ where D is a diagonal matrix

$$D = \text{diag}\{d_1, d_2, \ldots, d_m\}$$

If we restrict the matrix C so that its columns are orthogonal, that is $\mathbf{c}_i{}^T\mathbf{c}_j = 0$ if $i \neq j$, and also scale it so that $\mathbf{c}_i{}^T\mathbf{c}_i = 1$ for each i, we say that the columns of C are orthonormal, $C^T C = C C^T = I$ where I is the identity matrix with each diagonal element = 1 and each off-diagonal element = 0. Also, if $S = A^TA$ and $D = B^TB = C^TA^TAC$, S and D are said to be similar. Note that $D = C^TSC$ and $S = C^TDC$.

This is implicit in principal components in Chapter 4, Section 6, and is used in the construction of contrasts in Chapter 6.

A.3 SOME PROPERTIES OF MATRICES

We use the notation of Appendix A 2, Property 6. E and F are n×n matrices.

1. Transpose

(i) The transpose of the product FE is $E^T F^T$. Also $(E^T)^T = E$.
(ii) A matrix is symmetric if it equals its transpose. It is obvious that S and D are symmetric.

2. Determinant

We will not define the determinant here, but note that it indicates the size of the matrix. With a 3×3 matrix, the determinant indicates the volume enclosed by the vectors. Some properties of determinants are:

(i) $\det D = d_1 \times d_2 \times \cdots \times d_n$ where D is a diagonal matrix.
(ii) $\det EF = \det E \times \det F$.
(iii) With C defined above, $\det C = 1$ so that $\det D = \det C^TSC = \det S$. This is one implication of D and S being similar.
(iv) If $\det E = 0$, E is said to be singular and if $\det E \neq 0$, E is said to be nonsingular.
(v) If E is nonsingular, $\det E^{-1} = 1/\det E$.

4. Eigenvalues (or latent roots or characteristic roots)

(i) For the diagonal matrix D, the eigenvalues are the diagonal elements.
(ii) By definition, if λ is an eigenvalue of E then λ satisfies the equation $\det(A-\lambda I) = 0$.
(iii) The eigenvalues of S and of $T^{-1}ST$, for any T, are the same as the eigenvalues of D. This is another implication of S and D being similar.
(iv) The vectors $\mathbf{c}_i$ in C are called eigenvectors corresponding to each d_i.

5. Trace

(i) Definition: trace E = Σe_{ii}, the sum of the diagonal elements.
(ii) For the diagonal matrix D, trace D = Σd_i.
(iii) If E, F and G are conformable then trace EFG = trace GEF = trace FGE.
(iv) Trace S = trace C^T D C = trace D, another implication of D and S being similar.

6. Positive definite

(i) Definition: F is positive definite if $\mathbf{y}^T F \mathbf{y} > 0$ for all possible values of **y**.
(ii) D is positive definite if each $d_i > 0$ since
$$\mathbf{y}^T D \mathbf{y} = \sum d_i y_i^2$$
(iii) S is positive definite if D is positive definite, and vice versa, as

$$\mathbf{y}^T S \mathbf{y} = \mathbf{y}^T C D C^T \mathbf{Y} = \mathbf{z}^T D \mathbf{z}$$

If D is positive definite then $\mathbf{z}^T D \mathbf{z}$ is always positive and hence $\mathbf{y}^T S \mathbf{y}$ is always positive.

Positive definite sums of squares are important in statistics as they are the basis for all variance estimates.

7. Idempotent

(i) Definition: P is indempotent if it is symmetric and PP = P.
(ii) The eigenvalues of P are 0 or 1.
(iii) Trace P = rank P.
(iv) I-P is idempotent.

APPENDIX B
EXPECTATION, LINEAR AND QUADRATIC FORMS

B.1 EXPECTATION

The expectation of a random variable is the population mean of the random variable. The expectation operator is linear which is formally stated as

1. If a random variable y has expected value μ then $E(ay + b) = a\mu + b$, where a and b are constants.

2. A function f( ) of the random variable y has expectation $E[f(y)] = f(\mu)$ if f( ) is a linear function, but otherwise the equality may not hold.

The prediction from a linear model is the expectation of a linear function of the estimates $\hat{\beta}$.

B.2 LINEAR FORMS

1. If the random variables, y_1 and y_2 have means μ_1 and μ_2, and variances $\sigma_1{}^2$ and $\sigma_2{}^2$, and covariance $(y_1, y_2) = \sigma_{12}$, then for the linear combination $a_1y_1 + a_2y_2$

$$\text{Mean} = a_1\mu_1 + a_2\mu_2$$

$$\text{Variance} = a_1\sigma_1^2 + a_2\sigma_2^2 + 2a_1a_2\sigma_{12}$$

This can be generalized to n variables:

$\sum a_i y_i$ has mean $\sum a_i \mu_i$

and variance $\sum a_i^2 \sigma_i^2 + \sum_{i \neq j} a_i a_j \sigma_{ij}$

2. If the y_i are normally distributed then so is the sum, $\Sigma a_i y_i$. The above can be written in vector notation as:
If $\mathbf{y} \sim N(\boldsymbol{\mu}, V)$ then $\mathbf{a}^T\mathbf{y} \sim N(\mathbf{a}^T\boldsymbol{\mu}, \mathbf{a}^T V \mathbf{a})$ where V is the variance-covariance matrix having diagonal elements σ_i^2 and off-diagonal elements σ_{ij}.

B.3 QUADRATIC FORMS

1. If $\mathbf{y} \sim N(\mathbf{0}, \sigma^2 I)$, then $\mathbf{y}^T A\, \mathbf{y}/\sigma^2 \sim \chi(m)$ if A is idempotent of rank m. Recall that A is idempotent means that $A^T = A$ and $AA = A$.

If $A = I-P$, this result gives the distribution of $SSE = \mathbf{y}^T(I-P)\mathbf{y}$

2. If $\mathbf{y} \sim N(\mathbf{0}, \sigma^2 I)$ then the two quadratic forms $\mathbf{y}^T A \mathbf{y}$ and $\mathbf{y}^T B \mathbf{y}$ are independently distributed if $AB = 0$.

If the matrix of predictor variables X can be split into two groups of orthogonal vectors X_1 and X_2, the corresponding projection matrices P_1 and P_2 satisfy $P_1 P_2 = 0$. This result therefore shows that SSR_1 and SSR_2 are independent.

3. If $\mathbf{y} \sim N(\mu, V)$ then the expectation of the quadratic form $\mathbf{y}^T A \mathbf{y}$ is given by $E(\mathbf{y}^T A\, \mathbf{y}) = \text{trace}\,(AV) + \mu^T A\, \mu$.

A could be the matrix I-P, in which case this result gives the expected increase in SSE when the model is not the correct one.

B.4 THE F-STATISTICS

This is a ratio of two independent estimates of variance. If A and B are independent and idempotent with rank p, q respectively and $y \sim N(\mathbf{0}, \sigma^2 I)$ then $(\mathbf{y}^T A \mathbf{y}/p)/(\mathbf{y}^T B \mathbf{y}/q)$ has an $F_{p,q}$ distribution.

This result gives the basis for the overall F-test for the model by taking $P = A$ and $I-P = B$.

APPENDIX C
DATA SETS

In this appendix, we display the data sets which have been used as illustrations in this book.

C.1 ULTRA-SOUND MEASUREMENTS OF HORSES' HEARTS

The following data consists of ultra-sound readings on the hearts of 49 horses. Columns 1, 3, and 5 are the ultra-sound measurements during systole phase of the inner wall, outer wall and exterior width in millimetres; columns 2, 4, and 6 are measurements of the same variables during the diastole phase. The animals all suffered from certain conditions which required them to be killed and subsequently their hearts were weighed and these weights in kilograms are shown in the last column.

ID	Inner wall		Outer wall		Exterior		Weight
1	3.8	1.9	2.4	1.5	10.8	10.0	1.432
2	3.0	1.7	2.8	1.7	11.6	12.0	1.226
3	2.9	1.9	2.4	1.7	12.8	12.8	1.460
4	3.6	2.0	2.5	1.7	13.5	13.6	1.354
5	4.3	2.8	2.7	2.0	14.0	14.0	2.211
6	3.6	2.3	2.8	1.7	12.7	13.1	1.212
7	3.8	2.5	2.9	2.0	15.9	15.2	1.800
8	3.7	2.7	3.5	2.5	14.7	16.2	1.758
9	3.3	2.2	3.0	1.8	14.0	14.3	1.701
10	3.6	2.6	2.6	2.0	11.5	12.4	1.510

11	3.6	2.3	2.7	2.2	11.7	13.8	1.914
12	4.4	2.9	3.6	2.6	18.1	18.6	2.998
13	3.3	2.3	3.0	2.1	14.5	16.8	2.261
14	3.8	2.5	2.3	1.5	15.2	16.2	2.240
15	3.8	2.6	3.3	2.6	15.4	16.8	1.800
16	2.4	1.4	2.0	1.3	8.6	8.9	1.276
17	2.7	2.0	2.4	1.8	11.7	12.5	1.517
18	5.3	3.8	3.8	2.2	19.0	18.9	3.296
19	3.4	2.3	2.6	2.0	14.6	15.1	1.878
20	3.3	2.4	2.4	1.6	11.1	11.9	1.751
21	3.6	2.1	2.3	1.7	13.0	14.1	1.396
22	3.1	1.9	2.2	1.7	11.9	13.2	1.772
23	3.9	2.3	3.1	2.2	15.7	16.7	2.998
24	3.0	1.8	2.3	1.3	9.4	9.8	1.205
25	2.3	1.7	1.5	1.2	10.9	12.2	1.637
26	3.3	2.4	2.8	1.6	14.0	14.4	1.311
27	2.5	1.3	2.3	1.2	8.4	9.7	1.091
28	2.2	1.9	2.1	1.5	10.9	11.5	1.155
29	2.7	2.0	2.5	1.6	11.7	12.3	1.035
30	2.6	1.4	2.3	1.4	10.4	10.0	0.999
31	2.9	2.2	2.4	1.4	14.0	14.0	1.559
32	2.7	2.2	1.8	1.4	11.3	11.9	1.481
33	3.3	2.3	2.8	2.1	10.7	12.7	1.658
34	1.7	1.3	1.5	1.0	7.1	8.1	1.500
35	4.6	3.7	4.3	2.4	17.2	18.7	4.683
36	3.3	2.2	2.3	1.5	10.6	11.8	1.662
37	5.1	3.8	4.0	3.0	17.9	19.9	4.241
38	4.7	3.2	3.8	2.3	15.9	17.2	4.572
39	4.8	3.2	3.4	2.1	13.0	15.0	4.100
40	3.9	3.1	3.8	2.6	14.8	17.2	3.720
41	5.3	4.0	3.7	2.1	16.0	17.5	4.315
42	4.7	3.3	3.5	2.2	16.9	18.0	4.320
43	3.5	2.5	3.8	2.2	13.0	15.6	3.390
44	3.4	2.2	3.1	2.3	13.4	13.9	4.010
45	4.0	2.1	4.4	3.0	15.1	16.0	2.970
46	5.3	3.3	4.8	3.3	14.8	15.7	3.431

C.2 PH MEASUREMENTS OF LEAF PROTEIN

Leaf protein concentrate (protein extracted from grass) has been shown biogically to contain a factor which affects protein digestion. In vitro enzyme assays were carried out to help characterize this factor by measuring the pH drop with time due to the action of trypsin enzyme on standard casein. The experiment was carried out three times yielding the following three sets of results.

Time,x	pH values,y			Log of pH, ln y			ln x
1	8.00	8.00	8.00	2.07944	2.07944	2.07944	0.00000
2	7.74	7.76	7.74	2.04640	2.04898	2.04640	0.69315
3	7.61	7.63	7.61	2.02946	2.03209	2.02946	1.09861
4	7.53	7.54	7.52	2.01889	2.02022	2.01757	1.38629
5	7.47	7.49	7.45	2.01089	2.01357	2.00821	1.60944
6	7.42	7.45	7.40	2.00418	2.00821	2.00148	1.79176
7	7.38	7.41	7.36	1.99877	2.00283	1.99606	1.94591
8	7.35	7.38	7.30	1.99470	1.99877	1.98787	2.07944
9	7.33	7.36	7.28	1.99197	1.99606	1.98513	2.19722
10	7.30	7.33	7.25	1.98787	1.99197	1.98100	2.30258
11	7.28	7.32	7.23	1.98513	1.99061	1.97824	2.39789

C.3 LACTATION RECORDS OF COWS

The following data shows the quantity of milk in units of 0.5 litres in a 24 hour period which was delivered on one day a week for 38 weeks by 5 cows.

Week #	Cow 1	Cow 2	Cow 3	Cow 4	Cow 5
1	15.23	24.10	24.82	19.34	5.28
2	15.23	24.10	24.82	19.34	5.28
3	16.30	20.15	29.20	20.74	10.61
4	15.34	28.16	26.22	24.07	11.03
5	14.86	29.59	25.86	22.79	11.04
6	16.20	27.11	24.45	23.37	10.52
7	13.49	31.27	21.80	25.20	11.08
8	15.25	27.46	22.51	23.24	10.58
9	14.94	28.05	23.21	22.18	11.92
10	15.55	29.62	22.38	21.34	10.29
11	16.20	25.01	20.25	21.28	9.69
12	15.08	22.30	22.64	20.81	10.74
13	15.23	25.50	20.78	20.08	11.02
14	11.59	25.74	18.04	19.33	10.29
15	13.00	22.45	19.00	17.32	10.07
16	11.51	22.44	17.70	16.44	10.72
17	12.34	21.18	17.47	18.60	9.75
18	12.31	21.57	19.01	17.36	10.20
19	12.51	20.69	17.68	17.09	9.14
20	11.91	19.96	18.18	17.17	9.17
21	11.70	21.85	18.10	16.80	8.46
22	12.03	20.97	17.93	17.90	8.79
23	11.08	21.96	18.42	18.14	9.48
24	10.54	20.87	19.06	17.53	8.18
25	11.56	20.87	17.25	15.11	9.34
26	10.97	19.75	17.36	14.90	8.33
27	10.55	19.90	18.02	13.15	8.59
28	9.14	17.79	15.04	13.28	6.97
29	8.89	16.83	17.02	12.69	6.21

30	7.73	19.31	15.19	11.30	5.53
31	7.10	16.49	15.11	12.87	3.90
32	8.09	15.39	16.02	11.20	6.40
33	7.64	16.13	13.28	11.39	4.07
34	9.06	13.36	13.59	10.39	4.41
35	6.60	12.00	14.21	10.45	3.29
36	6.86	12.87	13.02	9.45	2.63
37	6.22	12.07	10.61	8.88	3.48
38	6.31	11.50	13.37	8.04	2.18

C.4 SPORTS CARS

The following data consists of information given in advertisements to sell a certain kind of automobile, namely the British built MGBT. The prices asked for these cars are shown in column 4, the year of manufacture in column 3, the week of the year in which the advertisement appeared, and the final column has a code of 1 if the seller was an individual and a code of 2 if the seller was a used car firm.

ID	Week	Year	Price	Seller
1	7	67	6100	1
2	7	67	6000	1
3	7	73	8800	1
4	7	77	13995	2
5	8	67	6000	1
6	12	67	8400	1
7	12	68	6800	1
8	13	67	6795	2
9	13	68	6990	2
10	13	72	8495	2
11	13	71	7995	2
12	13	70	7995	2
13	16	72	8495	2
14	16	78	15995	2
15	19	77	15995	2
16	20	67	6200	1
17	21	72	7800	1
18	23	67	6500	1
19	23	78	18000	1
20	31	70	7500	1
21	31	78	17500	1
22	32	79	16990	2
23	49	70	7800	1
24	51	71	8000	1
25	55	71	9500	1

26	55	76	13995	2
27	62	67	5500	1
28	61	68	6500	1
29	61	70	8500	1
30	58	72	6500	1
31	56	77	15995	2
32	49	70	7800	1
33	51	74	10495	2
34	54	77	15995	2
35	64	68	6990	2
36	64	75	13995	2
37	64	77	16500	1

C.5 HOUSE PRICE DATA

In New Zealand, each residential section and house is valued by government valuors. In this data set we list the selling prices of a sample of houses sold in Palmerston North in 1982. Also shown is the government valuations of land and of the house along with the size of the house (in units of 10 square metres whhich is approximately 100 square feet) and the area of the section in hectares (1 hectare is approximately 2.5 acres).

ID	Price	GV-House	GV-Land	Size	Section
1	48000	14700	8800	10	1.0700
2	37000	14100	8400	10	0.0635
3	29000	8900	8600	10	0.0573
4	66000	30600	15400	20	0.0645
5	66000	20700	8800	10	0.0861
6	39500	15800	10200	9	0.0849
7	73000	31300	11700	15	0.0743
8	45000	12000	10500	12	0.0654
9	36000	5900	10600	12	0.0607
10	75000	28600	10900	17	0.1353
11	85000	33600	12900	16	0.0688
12	67000	30000	9000	18	0.0716
13	61000	14300	15700	11	0.0850
14	56300	10100	9900	15	0.0612
15	46000	16600	8900	10	0.0607
16	54000	16700	7300	9	0.0692
17	74000	21500	9500	10	0.0650
18	49000	17400	8100	11	0.0670
19	114000	35900	27600	18	0.2023
20	52000	13600	8400	8	0.0721
21	66750	26200	8800	13	0.0738
22	43500	14800	9200	9	0.0674

23	51000	17400	8600	11	0.0718
24	40000	12900	8100	9	0.0774
25	46400	18000	9000	10	0.0759
26	58500	22200	8300	13	0.0682
27	40000	11500	7500	9	0.0692
28	38600	10800	7700	10	0.0676
29	48000	14700	9300	9	0.0674
30	41200	12000	9500	11	0.0700
31	54500	19200	9300	11	0.0737
32	40000	12300	11200	8	0.0772
33	58000	18300	13200	11	0.0794
34	52500	16600	10900	10	0.0631
35	45000	15200	7800	9	0.0691
36	44000	16700	7800	10	0.0639
37	21000	5900	7100	11	0.0539
38	100000	40500	22000	19	0.1206
39	64000	8600	12900	12	0.0819
40	56500	16500	14500	12	0.0999
41	53000	20500	9500	14	0.0642
42	36000	11100	7900	12	0.0617
43	40000	13800	9700	9	0.0645
44	55000	17700	7800	13	0.0675
45	45000	16600	7900	10	0.0675
46	84000	27900	11600	16	0.0936
47	45500	16000	9000	12	0.0727
48	63000	26100	3400	14	0.0920
49	37500	2600	9400	12	0.1104

C.6 COMPUTER TEACHING DATA

The following experiment was testing the effectiveness of using a computer to teach word recognition to handicapped children. The factors compared in the experiment were computer against human, and a word focus method against a word pairing method. There were eight words in each list, four items of food and four tool names, and whether the food items or the tools names were presented first was a third factor. The subjects were shown eight words and were assessed by the number they could correctly name.

Subject	Teacher	Method	Order	PreTest	PostTest
1	Computer	Pairing	Fd_Tl	0	2
2	Computer	Pairing	Tl_Fd	0	1
3	Computer	Pairing	Fd_Tl	5	3
4	Human	Focus	Fd_Tl	4	8
5	Computer	Pairing	Tl_Fd	0	0

6	Computer	Focus	Fd_Tl	3	5
7	Human	Pairing	Tl_Fd	2	3
8	Computer	Pairing	Fd_Tl	6	8
9	Computer	Focus	Tl_Fd	3	5
10	Human	Pairing	Tl_Fd	2	1
11	Computer	Focus	Tl_Fd	0	0
12	Human	Pairing	Tl_Fd	1	7
13	Human	Focus	Fd_Tl	0	5
14	Human	Pairing	Fd_Tl	5	8
15	Computer	Pairing	Fd_Tl	2	5
16	Human	Focus	Fd_Tl	1	6
17	Human	Focus	Tl_Fd	0	0
18	Computer	Focus	Tl_Fd	2	0
19	Computer	Pairing	Tl_Fd	0	2
20	Computer	Pairing	Tl_Fd	0	4
21	Computer	Focus	Fd_Tl	1	0
22	Human	Pairing	Fd_Tl	2	4
23	Human	Focus	Fd_Tl	2	4
24	Computer	Focus	Fd_Tl	0	3
25	Computer	Pairing	Tl_Fd	2	0
26	Human	Pairing	Fd_Tl	0	4
27	Human	Pairing	Fd_Tl	1	8
28	Human	Pairing	Fd_Tl	0	1
29	Human	Focus	Fd_Tl	2	1
30	Human	Focus	Fd_Tl	2	4
31	Computer	Focus	Tl_Fd	6	7
32	Computer	Pairing	Tl_Fd	4	0
33	Computer	Pairing	Tl_Fd	0	1
34	Human	Pairing	Tl_Fd	4	8
35	Computer	Focus	Fd_Tl	0	1
36	Human	Focus	Fd_Tl	6	7
37	Computer	Pairing	Fd_Tl	0	4
38	Human	Pairing	Tl_Fd	2	5
39	Computer	Pairing	Fd_Tl	1	0
40	Computer	Focus	Fd_Tl	6	8
41	Computer	Focus	Tl_Fd	1	2
42	Human	Focus	Tl_Fd	2	0
43	Human	Focus	Fd_Tl	0	0
44	Human	Focus	Tl_Fd	1	3
45	Human	Focus	Tl_Fd	5	8
46	Computer	Focus	Fd_Tl	0	5
47	Human	Pairing	Fd_Tl	8	8
48	Computer	Focus	Tl_Fd	2	0
49	Human	Pairing	Tl_Fd	0	2

C.7 WEEDICIDE DATA

An experiment to test a new weedicide at several rates and in various combinations with existing weedicides had the following treatments:

A: Control
B: 0.5 l X
C: 1.0 l X
D: 1.5 l X
E: 1.0 l X + Igan
F: 1.0 l X + M C P B
G: 1.0 l X + Dinosob
H: 1.0 l X + 1.0 l Trietazine
I: 1.0 l X + 2.0 l Trietazine
J: 2.0 l Trietazine
K: Hoegraes

The experiment was arried out on 44 plots of peas as a complete randomized block experiment with pea yields as given below:

Treatment	Rep 1	Rep 2	Rep 3	Rep 4	Mean
A	3886	3023	2543	3167	3155
B	3730	3679	3326	3557	3573
C	3427	3713	3564	3357	3515
D	3651	3953	3418	3606	3657
E	3580	3143	3420	3070	3303
F	3808	3953	3556	3228	3636
G	3292	3621	3287	3544	3436
H	3770	3777	3488	3560	3649
I	3608	3668	3731	3123	3533
J	3866	4079	3040	2986	3493
K	4124	3663	3547	3360	3674

s2 = 68627 with 30 degrees of freedom.

References

Draper, N.R. and Smith, H. (1981). Applied Regression Analysis, (2nd edition). Wiley:New York.

Edwards, H.P., (1984). RANKSEL: A ranking and selection package. American Statistician, 38: 158-159.

Freedman, D., Pisani, R., and Purves, R. (1978). Statistics. Norton:New York.

Fisher, R.A. and Yates, F. (1974). Statistical Tables for Biological, Agricultural and Medical Research, (6th edition). Longman:London.

Gibbons, J.O., Olkin, I., Sobel M. (1977). Selecting and Ordering Populations: A New Statistical Methodology. Wiley, New York.

Hoaglin, D.C. and Welsch, R.E. (1978). The Hat Matrix in regression and ANOVA. The American Statistician, 32: 17-22.

Hoerl, A.E. and Kennard, R.W. (1970). Ridge regression. Biased estimation for non-orthogonal problems. Technometrics, 12: 55-67.

Hoerl, A.E. and Kennard, R.W. (1981). Ridge regression-1980 Advances, algorithms, and applications. American Journal of Mathematical and Management Sciences, 1: 5-83.

Hogg, R.V. (1974). Adaptive robust procedures: a partial review and some suggestions for future applications and theory. Journal of the American Statistical Association, 69: 909-925.

John, P.W.M. (1971). Statistical Design and Analysis of Experiments. Macmillan, New York.

Pearson, E.S. and Hartley H.O. (1976). Biometrika Tables for Statisticians, Vol. 1, (3rd edition). Griffin

Speed, F.M. and Hocking R.R. (1976). The use of R( )-notation with unbalanced data. American Statistician, 30 : 30-33.

"Student", (1937). Comparison between balanced and random arrangements of field plots. Biometrika, 29 : 363-379

Wewala, G.S. (1980). Design and analysis of mixture experiments. Unpublished M.Sc. thesis, Massey University, New Zealand.

INDEX

- I -

- J -

- L -

- M -

- N -

- W -